14

TESI

THESES

tesi di perfezionamento in Matematica sostenuta il 20 settembre 2007

Giuseppe Della Sala
Fakultät für Mathematik
University of Vienna
Nordbergstraße 15
1090 Wien, Austria

Geometric Properties of Non-compact CR Manifolds

Giuseppe Della Sala

Geometric Properties of Non-compact *CR* Manifolds

EDIZIONI
DELLA
NORMALE

ISBN: 978-88-7642-348-2

Contents

Introduction

This Ph.D. thesis is devoted to the study of several problems arising in CR geometry, most of which regarding non-compact manifolds embedded in $\mathbb{C}^n$. Sometimes, the non-compact manifold will be an unbounded, but closed submanifold of $\mathbb{C}^n$: this is the case for Chapters 2, 6, 7 and for Section 4.2. Other instances of non-compactness will appear considering bounded manifolds, which are not necessarily closed in $\mathbb{C}^n$. Problems of this kind are treated in Chapter 3 as well as in some parts of Chapter 7.

The basic notion in CR geometry is that of CR function. Let S be a hypersurface of $\mathbb{C}^n$, $n \geq 2$, and let $f : S \to \mathbb{C}$ be a function of class C^1. We say that f is a CR (Cauchy-Riemann) function if it satisfies the following condition:

$$df \wedge (dz_1 \wedge \cdots \wedge dz_n)|_S = 0. \qquad (0.0.1)$$

Condition (0.0.1) was written for the first time by Severi (see [42, 56]) who studied the problem of characterization of traces on real hypersurfaces of holomorphic functions. He proved that condition (0.0.1) is necessary and sufficient in order for an analytic function to be the trace of a function holomorphic on a neighborhood. Condition (0.0.1) is obviously also necessary for regular boundary values of holomorphic functions on a bounded domain $\Omega \subset \mathbb{C}^n$ (see Section 1.4 and [12, 47]). However, the local extension problem of CR functions (*i.e.* f is a CR function defined on a hypersurface $M \subset U$, where U is a domain of $\mathbb{C}^n$) is not always solvable. It is easy to see that the situation is more complicated, and in fact the local (Levi) convexity properties of M play a role. The first important result in this vein was proved by Lewy [42] for $n = 2$, under the hypothesis that M is strongly Levi convex (see also [43]). Afterwards, the extension problem has been widely treated. We refer to Section 1.4 for a brief overview of the subject.

The functions satisfying the tangential CR condition have, since then, given rise to a whole category of CR geometric objects. The simplest example is perhaps the following: let $M \subset \mathbb{C}^n$, $n \geq 2$, be a real hypersurface and let $f : M \to \mathbb{C}$ be a CR function. Let N be the graph of f; then N is a real, 3-codimensional submanifold of $\mathbb{C}^{n+1}$. If $p \in N$ and J is the complex structure on $\mathbb{C}^{n+1}$, we define the *complex tangent space* $H_p(N)$ as

$$H_p(N) = T_p(N) \cap JT_p(N)$$

which is the greatest complex subspace of $T_p(N)$. It turns out that, f being a CR function,

$$\dim_{\mathbb{C}} H_p(N) = n - 1 \text{ for all } p \in N;$$

we say that N is *maximally complex*. Clearly, this is a non-generic (in some sense, the *most* non-generic) situation for a submanifold of $\mathbb{C}^{n+1}$ and, indeed, it is easily seen that it is a necessary condition for N to be the boundary of a complex submanifold (as expected, since the CR condition is a necessary one for f to be the trace of a holomorphic function). This fact leads to a natural question: given a maximally complex submanifold on $\mathbb{C}^n$, is it the boundary of a complex manifold? An answer when $M = \gamma$ is a compact curve was given by Wermer [66]; however, in this case the maximal complexity condition is trivial, and the right requirement is an integral condition (hence inherently global) which is also linked to the polynomial approximation on γ. The case of a compact, maximally complex submanifold of $\mathbb{C}^n$ with real dimension ≥ 3 was solved by Harvey and Lawson [29], in terms of holomorphic chains rather than manifolds because, of course, singularities cannot be avoided (although in some special case it can be proved that they are isolated). In Chapters 2 and 3 we deal with some variants of the boundary problem for complex submanifolds, by dropping in different ways the compactness hypothesis but still imposing suitable geometric constraints (thus allowing to treat the problem by means of Harvey-Lawson's result).

The maximally complex submanifolds are elements of a wider class of *CR submanifolds*, characterized by the fact that their complex tangent spaces have constant dimension and therefore form a subbundle of $T(N)$. On CR submanifolds it is possible to carry out a somewhat similar analysis to the usual one on $\mathbb{C}^n$; for example, a *Cauchy-Riemann complex* is defined, with a $\overline{\partial}_b$ operator whose kernel is the space of CR functions, and *Dolbeault decompositions* are also defined for differential forms (or fields, currents etc.). In Chapter 1 we present a - very short - introduction to the basics of CR geometry which is sufficient for our scope, referring to [13] for a more complete account.

Another aspect of CR geometry that arises in a number of different situations is that of *Levi convexity*. Actually, this is a concept that pre-dates the *CR* category since it was introduced as early as 1910, when E. E. Levi [41] proved that it is a necessary condition for a domain $D \subset \mathbb{C}^2$ with smooth boundary to be a domain of holomorphy. Levi expressed, in real variables, a non-linear partial derivatives condition that has to be satisfied by the local defining function ρ of the boundary bD of D. Namely, the *Levi form*

$$\mathcal{L}(\rho)(\xi, \eta) = \sum_{i,j=1}^{n} \frac{\partial^2 \rho}{\partial z_i \partial \bar{z}_j} \xi_i \bar{\eta}_j,$$

where $\xi, \eta \in T^{1,0}(\mathbb{C}^n)$, must be positive (semi)-definite when restricted to $H(bD)$, *i.e. D* is *Levi convex*. It turns out that Levi convexity, which is a local condition on bD, gives in fact a characterization of domains of holomorphy (see [36]).

The Levi form $\mathcal{L}(N)$ can be also defined for the abstract *CR* manifolds; we give this definition, along with some elementary properties, in Section 1.3. If $\mathcal{L}(N)$ vanishes identically we say that N is *Levi flat*; in this case $H(N)$ is integrable and N is foliated by holomorphic submanifolds. Due to their "critical" nature, Levi flat manifolds emerge as "limit case" in various situations. For example, the extension of *CR* function from an hypersurface of $\mathbb{C}^n$ his generally possible up to a Levi flat hypersurface (see Theorem 1.4.7). Another example is the construction of the polynomial hull of a graph $S' \subset \mathbb{C}^2$ over the 2-sphere $S^2 \subset \mathbb{C} \times \mathbb{R}$, which turns out to be a Levi flat 3-manifold [57]. The first result in this direction was obtained by Bedford and Gaveau (see [9]), and a rather vast amount of research has been performed since then on the boundary problem for Levi flat hypersurfaces, mainly in $\mathbb{C}^2$. In Chapter 4 we describe some recent developments in $\mathbb{C}^n$ for $n \geq 3$ (contained in [22]), and we obtain some related results.

The study of geometric properties of Levi flat manifolds gives rise to a variety of very interesting problems (see [8, 14]). In Chapters 6 and 7, we deal with some of them in the case of Levi flat hypersurfaces in $\mathbb{C}^n$. As a result of a preliminary exploration, we show that if a Levi flat manifold is bounded in some directions, then its foliation is "trivial".

We pass on now to a more detailed description of the contents of the thesis.

In Chapter 1, we give a rapid introduction to those notions of *CR* geometry that will be employed in the rest of the thesis. We start by defining *CR* manifolds in an abstract way. Although abstract *CR* manifolds are the source of many interesting questions (for example, the embedding problem) we will be interested exclusively on embedded ones, so we shall mainly focus on them. Next, we introduce the *CR* condition for submanifolds of $\mathbb{C}^n$ (observing that they are also *CR* in the abstract sense) and we define the basic notations (holomorphic tangent space, *CR* dimension, genericity etc.). In the following section, we give the notion of *CR* application in an intrinsic way, though we later introduce the Cauchy-Riemann complex and the $\overline{\partial}_b$ operator only for embedded manifolds. Then, we provide a definition of the Levi form by means of brackets - valid also for abstract *CR* manifolds - that will be useful in Section 2.3. Levi flatness of *CR* manifolds *M* is characterized by the vanishing of $\mathcal{L}(M)$, then in view of Frobenius and Newlander-Nirenberg Theorems *M* is foliated by complex submanifolds.

Finally, we give a list of the principal results related to the extension of *CR* functions. Some are directly applied later (*e.g.* Lewy's Theorem, or Plemelj's formula), mainly in Chapters 2, 3 and 4. Others have been "inspiring examples", because of their formulation and the methods of proofs employed; this is the case, for example, for the proof of Lewy's Theorem and for the hypothesis (⋆) introduced by Lupacciolu (see Section 1.4).

Chapters 2, 3, 6, 7 and Sections 4.2, 4.3 contain the original results of the thesis.

Chapters 2 and 3 contain the results of joint works with A.Saracco [16, 17]. They both deal with the boundary problem for non-compact complex submanifolds of $\mathbb{C}^n$.

In Chapter 2, our datum is a maximally complex submanifold $M \subset b\Omega$, where Ω is a (weakly) pseudoconvex, unbounded subdomain of $\mathbb{C}^n$. Our purpose is to find a complex submanifold *W* contained in Ω, with isolated singularities (in such a way that we can consider its boundary in a "geometric" way, rather than in the sense of currents), such that $bW = M$. We achieve the result in two steps: first of all, we show that there exist a local extension of *M* into Ω (to be able to prove that, we assume that the Levi form of $b\Omega$ has an appropriate number - depending on the dimension of *M* - of positive eigenvalues); this will later allow to show that the singularities are isolated. Then, we cut *M* with a family of complex hyperplanes *H* such that $H \cap b\Omega$ is compact; in order to make the proof more transparent, we first assume that Ω is convex (thus the choice of such a family is clearly possible) and then we show that Lupacciolu's hypothesis (⋆) allows, up to a suitable embedding, to find again a family of hyperplanes which intersects *M* in a compact subset. We apply Harvey-Lawson's Theorem to each slice: the hard part is to show that the collection of the complex varieties so obtained forms an analytic subvariety, and this is achieved by studying the behavior of suitable integral representation formulas.

In Chapter 3, we change the point of view slightly. In this case, *M* is embedded as a closed submanifold of a domain $A \Subset b\Omega$, where Ω

is a strongly pseudoconvex domain in $\mathbb{C}^n$. We want to find "how far" it is possible to extend M as a complex subvariety of Ω with isolated singularities, and of course we want the answer to depend only on A. We are able to prove that the extension is possible on the compact connected component $\tilde{A}$ of $\Omega \setminus \widehat{bA}$, where $\widehat{bA}$ is the hull of bA with respect to the functions holomorphic on a neighborhood of Ω. The proof is achieved by applying the same general scheme as in Chapter 2; however, we are no longer able to cut M by complex hyperplanes and the slice is performed by (regular) level sets of holomorphic functions. This fact adds several technical difficulties, first of all, the parameter of the level sets cannot be used as a coordinate as before, and we are forced to introduce a new variable in order let the method work. Then, due to the non-convexity of the slices, it is more involved to show that the solutions obtained by different cuts agree. The natural question of the maximality of $\tilde{A}$ has proved somewhat elusive, and has yet to be settled; nevertheless, in some very simple cases $\tilde{A}$ is maximal.

Levi flat hypersurfaces are particular CR manifolds which play a special role for extension in complex analysis: as already observed, the presence of Levi flat hypersurfaces in domains of $\mathbb{C}^n$ is very often an obstacle to extend analytic objects through. The description of their geometric features can be very difficult. In the rest of this thesis, we prove some results in this direction.

In the first section of Chapter 4 we give an exposition of [22], indulging in some details but just sketching most of the proofs. This paper deals with the boundary problem for Levi flat hypersurfaces in $\mathbb{C}^n$, $n \geq 3$. The target is a real $(2n - 2)$-submanifold $S \subset \mathbb{C}^n$ with finitely many complex points. S is assumed to be non-minimal in its CR points (i.e. it is required that S is not the minimal submanifold that integrates its own complex tangent bundle): this non-generic condition, trivial when $n = 2$, is a necessary one for S to be a local Levi flat boundary. Moreover it is assumed - in some ways, similarly to Bedford and Gaveau's work [9] - that all the complex points of S are elliptic. In this situation, S is showed to be actually a $(2n - 2)$-sphere with just two complex elliptic points p, q; moreover, $S \setminus \{p, q\}$ is foliated by compact, maximally complex CR-orbits homeomorphic to S^{2n-3}. An Harvey-Lawson Theorem with C^∞ parameter is applied to the family of these CR-orbits, and a solution of the problem is thus obtained as an immersed Levi flat variety with negligible singularities.

The following sections contain some results which are in various ways related to [22]. The first one is linked to the result in Chapter 2, i.e. we employ the methods of this chapter to obtain a similar Harvey-Lawson Theorem with C^∞ parameter for some class of unbounded maximally complex submanifolds (that is, the same class we treat in Chapter 2). Next, we discuss the problem of the regularity of the Levi flat variety obtained in [22], proving that it is indeed a Levi flat manifold in some simple situations (namely, when S is a graph): the main tool is Shcherbina's characterization [57] of the polynomial hull of a graph in $\mathbb{C}^2$. Admittedly, this result is not about non-compact submanifolds, although it could be developed in such a direction when the boundary problem for unbounded Levi flat submanifolds of $\mathbb{C}^n$, $n \geq 3$, is studied.

In the last chapters of the thesis we consider some structure problems regarding unbounded Levi flat submanifolds of $\mathbb{C}^n$. Our starting point was the following: given a Levi flat manifold $S \subset \mathbb{C}^n$ which is the graph of some smooth function $f : \mathbb{C}^{n-1} \times \mathbb{R} \to \mathbb{R}$ (here is the - rather loose - connection with Chapter 4), what can be said on S if f is bounded? This question led to several related ones; in the end, it turned out that some of these can be settled as quite simple consequences of known results in the theory of analytic multifunctions.

Thus, Chapter 5 is an introduction to the subject of analytic multifunctions. These objects, which where first introduced by Oka [49], are set-valued functions $\mathbb{C} \to \mathbf{k}(\mathbb{C})$ (where $\mathbf{k}(\mathbb{C})$ denote the subset of $\mathcal{P}(\mathbb{C})$ formed by the compact subsets of $\mathbb{C}$) which behave in some ways as analytic function; namely, according to Oka's definition, the complementary of their graph is pseudoconvex. This is not the only definition that has been given, and actually other characterizations of analyticity are more adapted to our purpose because they are suitable to generalizations (for example, to set-valued functions $\mathbb{C}^m \to \mathbf{k}(\mathbb{C}^n)$). The definition that we give in Chapter 5, and the subsequent treatment of the matter, is based on paper by Ransford [51] which develops the theory in a rather elegant, formal way. Since we are only interested in providing an overview of the properties of analytic multifunctions, we omit most of the proofs.

Chapter 6 is devoted to the problem mentioned above of a bounded Levi flat graph $S \subset \mathbb{C}^n$. It is easy to see that it suffices to treat the problem for $n = 2$. The result that we achieve is that S is "trivially" foliated by complex hyperplanes (parallel to $\mathbb{C}^{n-1}$). The original approach we used to deal with the question - an analysis performed on each single leaf - can be circumvented, after an appropriate rational change of coordinates, by an application of the Liouville Theorem for analytic multifunctions, as done in Chapter 7. However, our method allows to actually obtain a sharper result, *i.e.* if the graph S is foliated (in a suitably regular way) by not necessarily complex submanifolds, and a leaf Σ is in fact a complex one, then the triviality result holds for Σ. To prove this result, we proceed as follows. We first prove that the projection of Σ on $\mathbb{C}$ contains the projection of its cluster points on S. Afterwards, denoting by Ω the image of the projection of Σ on $\mathbb{C}$, we examine the features of Ω; in particular, we prove that Ω is simply connected and that $\Sigma \rightarrow \Omega$ is single-sheeted. Then, we prove that in fact $\Omega = \mathbb{C}$ (which leads easily to the proof of thesis). This is achieved by using the previous observation on cluster points, along with the fact that an harmonic function (conjugated to a bounded one) cannot explode on too large a subset of the boundary of its domain of definition. We conclude with a discussion of the C^0 case.

Finally, in Chapter 7 we treat other problems regarding foliated manifolds whose foliation includes complex leaves. First, we deal with the triviality of the foliation of a Levi flat submanifold S contained in a cylinder of the kind $\{|w| < c\}$, where w is a complex coordinate in $\mathbb{C}^n$. For the case $n = 2$ or, more generally, for the case of a hypersurface of $\mathbb{C}^n$, the result follows almost immediately by the Liouville Theorem for analytic multifunctions. Next, we formulate the problem also for a Levi flat submanifold S with codimension greater than 1. Again, the problem can be solved by applying the theory of analytic multifunctions, although the application is slightly more involved. In this situation indeed, since the complementary of S is no longer pseudoconvex, some additional care is required to prove that S is anyway an analytic multifunction (in the broader sense, *i.e.* according to the characterization by psh functions). Then we apply again the Liouville Theorem for multifunctions; however, also this step is a bit less trivial than before, since we need to use (at least, in our method of proof) the characterization of the polynomial hull of curves in $\mathbb{C}^n$ for $n \geq 2$ (see [60]).

Chapter 1
The *CR* category

1.1. *CR* manifolds

Let M be a real, smooth, n-dimensional manifold, and let

$$T^{\mathbb{C}}(M) = T(M) \otimes_{\mathbb{R}} \mathbb{C}$$

be the complexified tangent bundle.

Definition 1.1.1. We say that M is a *CR manifold* if there exists a complex subbundle $A \subset T^{\mathbb{C}}(M)$ such that

- $A \cap \overline{A} = \{0\}$;
- A is involutive, *i.e.* for every pair of sections $P, Q \in \Gamma(M, A)$ it holds $[P, Q] \in \Gamma(M, A)$.

If $\dim_{\mathbb{C}} A = l$, we say that M has *CR dimension l*.

If M is a *CR* manifold we set $H^{\mathbb{C}}(M) = A \oplus \overline{A}$ and

$$H(M) = \mathsf{Re}(H^{\mathbb{C}}(M)) = \{X \in T(M) : \exists Y \in T(M) \text{ s.t. } X + iY \in H^{\mathbb{C}}(M)\}.$$

An (almost) complex structure J on $H(M)$ is then defined, in such a way that the diagram

$$
\begin{array}{ccc}
A & \xrightarrow{\;\;\;i\;\;\;} & A \\
{\scriptstyle \mathsf{Re}}\downarrow & & \downarrow{\scriptstyle \mathsf{Re}} \\
H(M) & \xrightarrow{\;\;\;J\;\;\;} & H(M)
\end{array}
$$

commutes (here we denote by "i"the multiplication by i). Then, A and $\overline{A}$ correspond to the eigenspaces of the complexification $J^{\mathbb{C}}$ of J, relative to the eigenvalues i and $-i$ respectively. We can construct J explicitly as follows. Let $p \in M$, and let $v_1, \ldots, v_l$ be a basis of A_p on $\mathbb{C}$; then $\overline{v}_1, \ldots, \overline{v}_l$ is a basis of $\overline{A}_p$ and

$$\alpha_k = \frac{v_k + \overline{v}_k}{2}, \quad \beta_k = i\frac{v_k - \overline{v}_k}{2}, \quad k = 1, \ldots, l$$

is a basis of $H_p(M)$ as a vector space on $\mathbb{R}$. We define $J_p : H_p(M) \to H_p(M)$ in the following way:

$$J_p(\alpha_k) = \beta_k, \quad J(\beta_k) = -\alpha_k \ k = 1, \ldots, l.$$

With this definition, it is clear that $J_p^2 = -Id$ and, moreover,

$$J_p^{\mathbb{C}}(v_k) = J_p^{\mathbb{C}}\left(\frac{\alpha_k - i\beta_k}{2}\right) = \frac{\beta_k + i\alpha_k}{2} = iv_k, \ J_p^{\mathbb{C}}(\bar{v}_k) = -i\bar{v}_k$$

for $k = 1, \ldots, l$. It is easy to see that, starting with another basis $v_1', \ldots, v_l'$, we obtain the same J_p: this implies that J_p is in fact a well defined, smooth tensor $J : H(M) \to H(M)$. Moreover, with this definition of J we have a canonical, complex linear isomorphism between $(H(M), J)$ and A, given by

$$\Gamma(M, H(M)) \ni X \to X - iJX \in \Gamma(M, A).$$

A first, trivial example of a CR manifold is given by any manifold M and $A = \{0\}$; in this case $H(M) = \{0\}$, and we say that M is *totally real*. At the "opposite side of the spectrum" there are the complex manifolds. Indeed, if M is a complex manifold we can set $A = T^{1,0}(M)$; then A is involutive and we have $\bar{A} = T^{0,1}(M)$, $A \oplus \bar{A} = T^{\mathbb{C}}(M)$, *i.e.* $H(M) = T(M)$ and thus $H(M)$ has the maximal possible dimension.

Embedded *CR* manifolds

Consider a smooth, real, m-dimensional submanifold $M \subset \mathbb{C}^n$, and let $J : \mathbb{C}^n \to \mathbb{C}^n$ be the isomorphism induced by multiplication by i. For any $p \in M$, define

$$H_p(M) = T_p(M) \cap J(T_p(M));$$

then $H_p(M)$ is the maximal J-invariant (*i.e.* complex) subspace of $T_p(M)$, and is called the *holomorphic tangent space* of M at p. If $d = 2n - m$ is the real codimension of M, we have

$$2n - 2d \leq \dim_{\mathbb{R}} H_p(M) \leq 2n - d$$

hence

$$n - d \leq \dim_{\mathbb{C}} H_p(M) \leq 2\left[n - \frac{d}{2}\right] \tag{1.1.1}$$

where $[x]$ denotes the largest integer smaller than x. Therefore, if $d < n$ we have $H_p(M) \neq \{0\}$. However, there is no need for $\dim H_p(M)$ to be independent from p.

Example 1.1.2. Let $M = S^2 \subset \mathbb{C} \times \mathbb{R} \subset \mathbb{C}^2$ (with coordinates (z, w), $z = x + iy$, $w = u + iv$). Then, at $p_1 = (0, 1)$ and $p_2 = (0, -1)$ it holds $T_{p_1}(S^2) = T_{p_2}(S^2) = \mathbb{C}_z$, *i.e.* p_1 and p_2 are *complex points*. On the other hand, for $q \in S^2 \setminus \{p_1, p_2\}$ we have that $T_q(S^2)$ is not a complex subspace of $\mathbb{C}^2$ (since it is contained in $\mathbb{C}_z \times \mathbb{R}$ and it is not $\mathbb{C}_z$), hence $H_q(S^2) = \{0\}$. In general, for a generic compact 2-dimensional submanifold S of $\mathbb{C}^2$, the number and the type (namely, the orientation and the hyperbolicity or ellipticity) of the complex points is linked to the Euler characteristic $\chi(S)$, hence it is not possible to give an embedding of S^2 in $\mathbb{C}^2$ as a CR submanifold. It is instead possible to construct such an embedding for the torus: indeed, the Šilov boundary T of the bidisc $D_z \times D_w \subset \mathbb{C}^2$ is a totally real embedding of the torus. To see this, observe that T is invariant under the complex linear transformations of $\mathbb{C}^2$

$$R_{\theta,\phi}(z, w) = (e^{i\theta} z, e^{i\phi} w)$$

which, moreover, act transitively on T. This means that either the tangent space of T is a complex line at every point, or T is totally real; but in the first case T would be a compact complex submanifold of $\mathbb{C}^2$, a contradiction.

Definition 1.1.3. We say that M is an *embedded CR submanifold* if the dimension of $H_p(M)$ is constant over M.

In such a case, the collection $\{H_p(M)\}_{p \in M}$ is in fact a subbundle of $T(M)$, since the dependence of $T_p(M) \cap J(T_p(M))$ on $p \in M$ must then be smooth. We say that a CR submanifold M of codimension d is *generic* if the dimension of $H(M)$ is the least possible, that is, by (1.1.1)

$$\dim_{\mathbb{C}} H(M) = \max\{n - d, 0\}.$$

This is indeed the generic situation, *i.e.* every submanifold can be locally perturbed to be (locally) CR with the holomorphic tangent bundle of minimal dimension (this can by seen by standard transversality methods). If d is odd, we say that a CR submanifold M of real codimension d is *maximally complex* if its CR dimension is the maximal possible, that is

$$\dim_{\mathbb{R}} H(M) = 2n - d - 1 = \dim_{\mathbb{R}} T(M) - 1.$$

Of course, for $d > 1$ this is a highly non-generic situation, and as we will see later it is linked to the existence of a complex submanifold bounded by M.

If $d = 1$, *i.e.* M is a hypersurface of $\mathbb{C}^n$, then by (1.1.1) we see that M is automatically a CR submanifold (which is both generic and maximally complex). Of course, in this case M is locally the boundary of a complex manifold (namely, an open subset of $\mathbb{C}^n$).

The notion of embedded CR submanifold introduced by the previous definition gives an abstract CR manifold as the ones described in 1.1.1. To see this, define $H^{1,0}(M) \subset H^{\mathbb{C}}(M)$ to be the eigenspace of $J^{\mathbb{C}}$: $H^{\mathbb{C}}(M) \to H^{\mathbb{C}}(M)$ relative to the eigenvalue i, and let $H^{0,1}(M) = \overline{H^{1,0}(M)}$. Then

$$H^{1,0}(M) = T^{\mathbb{C}}(M) \cap T^{1,0}(\mathbb{C}^n)|_M.$$

Putting $A = H^{1,0}(M)$, the previous expression shows that A is an involutive subbundle of $T^{\mathbb{C}}(M)$ (since it is the intersection of two involutive ones). Moreover, $A \cap \overline{A} = \{0\}$ and $A \oplus \overline{A} = H^{\mathbb{C}}(M)$ because A and $\overline{A}$ are the eigenspaces of $J^{\mathbb{C}}$ relative to the eigenvalues i and $-i$. Thus M is a CR manifold according to Definition 1.1.1, and the holomorphic tangent bundle is the same $H(M)$ defined previously.

1.2. CR functions

Let M, N be (abstract) CR manifolds, and let J_M, J_N be the complex structures defined in the previous section on $H(M)$ and $H(N)$ respectively.

Definition 1.2.1. A smooth mapping $f : M \to N$ is called a CR application if $df(H(M)) \subset H(N)$ and, moreover, df is complex linear with respect to the J_M, J_N structures, *i.e.*

$$df_p \circ (J_M)_p = (J_N)_{f(p)} \circ df_p$$

or, equivalently,

$$df_p^{\mathbb{C}}(A_M)_p \subset (A_N)_{f(p)}$$

for all $p \in M$.

It is clear that the composition of CR applications is again a CR application; moreover, if M and N are in particular complex manifolds, the previous definition requires df to be complex linear in the usual sense *i.e.* f is holomorphic.

Now, let $M \subset \mathbb{C}^n$ be a d-codimensional embedded CR submanifold, and let $\{\rho_i = 0\}_{i=1,\dots,d}$ be local equations for M. We define a *Cauchy-Riemann complex*, analogous to the one defined on the complex manifolds, in the following way. Let $\bigwedge^{p,q} T^\star(\mathbb{C}^n)$ be the vector bundle on $\mathbb{C}^n$ whose sections are the (p, q)-forms. Consider the restriction $\bigwedge^{p,q} T^\star(\mathbb{C}^n)|_M$ to M, and let $I^{p,q}$ be locally defined as

$$I^{p,q} = \bigwedge^{p,q} T^\star(\mathbb{C}^n) \cap \langle \overline{\partial}\rho_1, \dots, \overline{\partial}\rho_d \rangle$$

where $\langle \overline{\partial}\rho_1, \ldots, \overline{\partial}\rho_d \rangle$ is the ideal of the exterior algebra of $T^\star(\mathbb{C}^n)$ generated by $\overline{\partial}\rho_1, \ldots, \overline{\partial}\rho_d$. We define $T^{p,q}(M)$ to be the orthogonal of $I^{p,q}|_M$ in $\bigwedge^{p,q} T^\star(\mathbb{C}^n)|_M$, and consider the projection

$$t_M : \bigwedge^{p,q} T^\star(\mathbb{C}^n)|_M \to T^{p,q}(M).$$

If f is a section of $\bigwedge^{p,q} T^\star(\mathbb{C}^n)|_M$, we say that $t_M f$ is its *tangential part*. If, for any domain $U \subset M$, we denote by $\mathcal{E}_M^{p,q}(U)$ the sections of $T^{p,q}(M)$ on U, we define the *tangential Cauchy-Riemann operator*

$$\overline{\partial}_M : \mathcal{E}_M^{p,q}(U) \to \mathcal{E}_M^{p,q+1}(U)$$

as follows. For any $f \in \mathcal{E}_M^{p,q}(U)$, we consider an extension $\tilde{f}$ to an open subset of $\mathbb{C}^n$, and define

$$\overline{\partial}_M f = t_M(\overline{\partial}\tilde{f}).$$

If $\tilde{f}'$ is another local extension of f, then the result is the same. Indeed, $\tilde{f} - \tilde{f}' \in I^{p,q}$ and $\overline{\partial}I^{p,q} \subset I^{p,q+1}$ since

$$\overline{\partial}(\alpha\rho + \beta \wedge \overline{\partial}\rho) = \rho(\overline{\partial}\alpha) + (\alpha + \overline{\partial}\beta) \wedge \overline{\partial}\rho$$

for $\alpha \in \mathcal{E}^{p,q}(\tilde{U})$, $\beta \in \mathcal{E}^{p,q-1}(\tilde{U})$ and $\rho \equiv 0$ on U.

By definition, $\overline{\partial}_M$ shares some of the properties of the usual $\overline{\partial}$-operator in $\mathbb{C}^n$, namely, $\overline{\partial}_M^2 = 0$ and

$$\overline{\partial}_M(f \wedge g) = (\overline{\partial}_M) \wedge g + (-1)^{p+q} f \wedge (\overline{\partial}_M g)$$

for $f \in \mathcal{E}^{p,q}$ and $g \in \mathcal{E}^{r,s}$.

Definition 1.2.2. Let M be a *CR* submanifold of $\mathbb{C}^n$ and let $f : M \to \mathbb{C}$ be a C^1 function. We say that f is a *CR function* if $\overline{\partial}_M f = 0$.

Equivalently, f is *CR* if $\overline{L}f = 0$ for all the local sections L of $T^{1,0}(M)$. Moreover, it is simple to check that, if $\{\rho_i\}_{i=1\ldots,d}$ is a set of locally defining functions for M, then f is *CR* if and only if

$$(\overline{\partial}\tilde{f} \wedge \overline{\partial}\rho_1 \wedge \ldots \wedge \overline{\partial}\rho_d)|_M = 0$$

for all the local extensions $\tilde{f}$ of f to a neighborhood of M in $\mathbb{C}^n$. In particular, if M is a hypersurface of $\mathbb{C}^n$, a function on M is *CR* if and only if its $\overline{\partial}$-gradient is a multiple of the $\overline{\partial}$-gradient of a local defining function of M. This characterization shows that the restriction of a holomorphic function f to a *CR* submanifold is a *CR* function, since it is itself an extension of $f|_M$ such that $\overline{\partial}f = 0$.

Remark 1.2.3. In the previous definition of *CR* function f is supposed to be C^1, but only the derivatives of f along certain directions of $H^{\mathbb{C}}(M)$ are in fact taken in consideration. In some situations, it is easy to extend this notion to even non-continuous functions; for example, in $E = \{v = 0\} \subset \mathbb{C}^2(z, w)$, $(z = x + iy, w = u + iv)$, we may say that any function $f : E \to \mathbb{C}$ which is holomorphic with respect to z for any fixed u is a *CR* function.

We say that $f : M \to \mathbb{C}^n$ is a *CR application* all its components are *CR* functions. It can be proved that this definition agrees with the one given in 1.2.1 for the abstract case (taking in $\mathbb{C}^n$ the *CR* structure induced by the complex structure J).

As already observed, the restriction of a holomorphic function to a *CR* manifold is a *CR* function. The converse is not always true: if, as in the previous remark, we consider $E = \{v = 0\}$ and $f : E \to \mathbb{C}$, f is a *CR* function if and only if it is holomorphic with respect to z; so, if f is - for example - not continuous with respect to u, it is not the restriction of a holomorphic function. A case in which the converse holds is when M is a generic, real analytic *CR* submanifold of dimension at least n and f is real analytic (see [62]). In general, the local extension of a *CR* function defined in a hypersurface of $\mathbb{C}^n$ may depend on the convexity properties of M. The notion introduced in the following section proved to be the "good one" to give account of the convexity behavior of a hypersurface (or, more generally, a *CR* submanifold) in $\mathbb{C}^n$.

1.3. The Levi form

Let M be a *CR* manifold, and, for all $p \in M$, denote by π_p the projection

$$\pi_p : T_p^{\mathbb{C}}(M) \to T_p^{\mathbb{C}}(M)/H_p^{\mathbb{C}}(M).$$

Definition 1.3.1. The *Levi form* of M at $p \in M$ is the map $\mathcal{L}_p : A_p \to T_p(M)/H_p(M)$ defined by

$$\mathcal{L}_p(X_p) = \frac{1}{2i}\pi_p([\overline{X}, X](p))$$

where $X_p \in A_p$ and X is a local section of A around p such that $X(p) = X_p$.

The previous definition is well-given. Indeed, if $\{L_i\}$ is a local frame for A around p and $X = \sum x_i L_i$ (where x_i are C^∞ complex-valued functions defined in a neighborhood of p) then an easy calculation shows that

$$\pi_p[\overline{X}, X](p) = \sum \overline{x}_j(p)x_k(p)[\overline{L}_j, L_k](p),$$

which is an expression depending only on the value of X at p. The Levi form is preserved by *CR* morphisms. Indeed, if $F : M \to N$ is a *CR* application, then $dF(H^{\mathbb{C}}(M)) \subset H^{\mathbb{C}}(N)$ which implies that dF defines a map

$$T^{\mathbb{C}}(M)/H^{\mathbb{C}}(M) \to T^{\mathbb{C}}(N)/H^{\mathbb{C}}(N).$$

Moreover, by definition of $\mathcal{L}$, it is clear that

$$dF_p \circ \mathcal{L}_p^M = \mathcal{L}_{F(p)}^N \circ dF_p.$$

The Levi form gives information about the integrability of the vector bundle $H(M)$. If $H(M)$ is in fact integrable, then it can be shown that M is foliated by complex submanifolds.

Definition 1.3.2. We say that a *CR* manifold M is *Levi flat* if its Levi form vanishes.

Proposition 1.3.3. *Let M be a CR manifold. Then the following properties are equivalent:*

1. *M is Levi flat;*
2. *the distribution $p \to (H_p(M), J_p)$ is integrable;*
3. *M is foliated by (immersed) complex submanifolds which integrate the distribution $H_p(M)$.*

Proof. If M is a Levi flat manifold, its holomorphic tangent bundle $H^{\mathbb{C}}(M)$ is involutive. In fact, we have for all the pairs (X, Y) of local sections of A

$$[X + Y, \overline{X} + \overline{Y}] = [X, \overline{X}] + [Y, \overline{Y}] + [Y, \overline{X}] + [X, \overline{Y}],$$

$$[X + iY, \overline{X} - i\overline{Y}] = [X, \overline{X}] + [Y, \overline{Y}] + i[Y, \overline{X}] - i[X, \overline{Y}]$$

which implies that $[X, \overline{Y}]$ is a section of $H^{\mathbb{C}}(M)$; using this, and the fact that $A, \overline{A}$ are involutive, it can be proven immediately that $H^{\mathbb{C}}(M)$ (hence $H(M)$) is involutive. Frobenius theorem then implies that $H(M)$ is integrable and M is foliated by immersed submanifolds whose tangent space is $H(M)$, *i.e.* $1 \Rightarrow 2$. If 2 holds, Newlander-Nirenberg theorem [48] shows that the leaves M are (immersed) complex submanifolds, that is 3. Finally, it is clear that $3 \Rightarrow 1$; in fact, this property is sometimes assumed as an alternative definition of Levi flatness. $\square$

If M is an embedded submanifold (or, in general, when a metric is chosen), the hermitian metric on M allows to identify $T^{\mathbb{C}}(M)/H^{\mathbb{C}}(M)$ with the orthogonal $X^{\mathbb{C}}(M)$ of $H^{\mathbb{C}}(M)$ in $T^{\mathbb{C}}(M)$; moreover, due to the normalization factor $1/2i$ we have $\overline{\mathcal{L}} = \mathcal{L}$, thus $\mathcal{L}$ takes values on the real part $X(M)$ of $X^{\mathbb{C}}(M)$; we say that $X(M)$ is the *totally real part* of $T(M)$. In particular, if M is maximally complex then $\dim_{\mathbb{R}} X(M) = 1$ and then $\mathcal{L}$ can be seen as a real valued form; this is, for instance, the case for an hypersurface of $\mathbb{C}^n$.

In such a case, it makes sense to speak about the signature of $\mathcal{L}$, which is invariant by biholomorphisms because of the behavior of the Levi form under CR maps. If $\mathcal{L}$ is semi-definite (positive or negative) we say that M is *pseudoconvex*. With this definition, the boundary of a pseudoconvex domain of $\mathbb{C}^n$ is pseudoconvex, as follows from the next paragraph.

We check that the definition of Levi form given in 1.3.1 is consistent with the usual notion of Levi form used for hypersurfaces in $\mathbb{C}^n$. Let $M \subset \mathbb{C}^n$ be defined by $\{\rho = 0\}$; then the differential $d\rho = \sum_k \rho_{x_k} \partial/\partial x_k + \sum_k \rho_{y_k} \partial/\partial y_k$ can be expressed as

$$d\rho = \frac{1}{2} \sum_{k=1}^{n} \rho_{z_k} \frac{\partial}{\partial \bar{z}_k} + \frac{1}{2} \sum_{k=1}^{n} \rho_{\bar{z}_k} \frac{\partial}{\partial z_k}.$$

Since M is a hypersurface, and the field

$$J d\rho = \frac{1}{2} \sum_{k=1}^{n} i\rho_{z_k} \frac{\partial}{\partial \bar{z}_k} + \frac{1}{2} \sum_{k=1}^{n} -i\rho_{\bar{z}_k} \frac{\partial}{\partial z_k}$$

is orthogonal to both $d\rho$ and $H(M)$, it is a section of $X(M)$. Now, let X be a section of $T^{1,0}(M)$, *i.e.*

$$X = \sum_{k=1}^{n} \xi_k \frac{\partial}{\partial z_k}, \quad X \perp d\rho.$$

The last condition can be written as

$$\langle X, d\rho \rangle = 0 \Leftrightarrow \sum_{k=1}^{n} \xi_k \rho_{z_k} = 0 \Rightarrow \sum_{j,k=1}^{n} \bar{\xi}_j \frac{\partial \xi_k}{\partial \bar{z}_j} \rho_{z_k} + \bar{\xi}_j \xi_k \rho_{z_k \bar{z}_j} = 0. \quad (1.3.1)$$

The bracket of the Definition 1.3.1 can then be written as

$$[\overline{X}, X] = \sum_{k=1}^{n} \left(\sum_{j=1}^{n} \bar{\xi}_j \frac{\partial \xi_k}{\partial \bar{z}_j} \right) \frac{\partial}{\partial z_k} - \sum_{k=1}^{n} \left(\sum_{j=1}^{n} \xi_j \frac{\partial \bar{\xi}_k}{\partial z_j} \right) \frac{\partial}{\partial \bar{z}_k},$$

and the projection to the totally real part can be expressed as scalar product:

$$\langle [\overline{X}, X], J d\rho \rangle = -i \sum_{j,k=1}^{n} \left(\bar{\xi}_j \frac{\partial \xi_k}{\partial \bar{z}_j} \rho_{z_k} + \xi_j \frac{\partial \bar{\xi}_k}{\partial z_j} \rho_{\bar{z}_j} \right)$$

and by (1.3.1)

$$\frac{1}{2i}\langle[\overline{X}, X], Jd\rho\rangle = -\frac{1}{2}\sum_{j,k=1}^{n}\left(\overline{\xi}_j\xi_k\rho_{z_k\overline{z}_j} + \overline{\xi}_k\xi_j\rho_{z_j\overline{z}_k}\right) = -\sum_{j,k=1}^{n}\rho_{z_j\overline{z}_k}\xi_j\overline{\xi}_k$$

which is, up to a sign, the usual Levi form of ρ applied to X.

1.4. Extension of *CR* functions

As observed in Section 1.2, the trace of a holomorphic function on a *CR* submanifold is a *CR* function. Several results have been achieved to show that, in some situations, a *CR* function can be in fact extended to a holomorphic function defined on an appropriate domain of $\mathbb{C}^n$.

Global extension

The following theorem is in some ways analogous to the classical Hartogs extension theorem:

Theorem 1.4.1. *Let $D \Subset \mathbb{C}^n$ be a domain with smooth, connected boundary bD, and let $f \in C^0(bD)$ be a CR function. Then there exists $F \in C^0(\overline{D}) \cap \mathcal{O}(D)$ such that $F|_{bD} = f$.*

It can be given a proof based on the solvability of the $\overline{\partial}$-equation with compact support in $\mathbb{C}^n$. However, the theorem is also valid when f satisfies the *CR* condition in a *weak form* (see [29]), i.e. if $f \in L^1(b\Omega)$ satisfies

$$\int_{b\Omega} f\varphi = 0$$

for all $(n, n-1)$ forms φ such that $\overline{\partial}\varphi = 0$.

The function F of Theorem 1.4.1 can also be expressed in an explicit way by means of an appropriate kernel. The following representation formula was discovered independently by Bochner [12] and Martinelli [47].

Definition 1.4.2. For $z \in \mathbb{C}^n$, the *Bochner-Martinelli* kernel is the following form defined on $\mathbb{C}^n \setminus \{z\}$:

$$K_{BM}(z, \zeta)$$
$$= \frac{(n-1)!}{(2\pi i)^n}\sum_{\alpha=1}^{n}\frac{(-1)^\alpha(\overline{\zeta}_\alpha - \overline{z}_\alpha)}{|z-\zeta|^{2n}}d\zeta_1 \wedge ... \wedge d\zeta_n \wedge d\overline{\zeta}_1 \wedge ... \wedge d\hat{\overline{\zeta}}_\alpha \wedge ... \wedge d\overline{\zeta}_n$$

where $d\hat{\overline{\zeta}}_\alpha$ means that the differential $d\overline{\zeta}_\alpha$ is missing.

Theorem 1.4.3. *Let D be a bounded domain in $\mathbb{C}^n$, with connected boundary of class C^1. Let f be holomorphic in D and continuous up to $\overline{D}$. Then, for every $z \in D$,*

$$f(z) = \int_{bD} f K_{BM}(z, \cdot).$$

Then the statement of Theorem 1.4.1 can be made precise by saying that F can be obtained by means of integration of the form $f K_{BM}$ over bD. The fact that the result is a holomorphic function depends on the fact that f satisfies the CR condition. In such a way, it can also be seen that the extension F is in the same class of regularity as f.

Plemelj formula

The Bochner-Martinelli kernel can be employed also to prove, given a CR function f defined in a domain of a hypersurface M, the local existence of two holomorphic functions (defined on the two local components of $U \setminus M$, $U \subset \mathbb{C}^n$) whose difference is f. This is analogous to what can be proved for a continuous complex valued function defined on a curve on $\mathbb{C}$; in that case, of course, the CR condition becomes trivial.

Let $U \subset \mathbb{C}^n$ be an open domain, and let $M \subset U$ be a closed hypersurface. Define the *Bochner-Martinelli transform* of $f \in C^1(M)$ in the following way:

$$F(z) = \int_M f K_{BM}(z, \cdot).$$

Moreover, we define the *Cauchy principal value* at $z \in M$ as the following limit (provided that it exists):

$$P.V. \int_M f K_{BM}(z, \cdot) = \lim_{\varepsilon \to 0^+} \int_{M \setminus B(z,\varepsilon)} f K_{BM}(z, \cdot).$$

With these definitions, we can state the following result:

Theorem 1.4.4. *Suppose that M divides U in exactly two connected components U^+, U^-, and that it is oriented in such a way that $dU^+ = M$. Let $f \in C_0^1(M)$ and let F be its Bochner-Martinelli transform, defined on $U \setminus M$. Then $F|_{U^\pm}$ has a continuous extension $F^\pm$ to $U^\pm \cup M$; these extensions satisfy the* Plemelj formula

$$F^\pm(z) = \pm \frac{1}{2} f(z) + P.V. \int_M f K_{BM}(z, \cdot)$$

for all $z \in M$. Moreover, for any compact subset $K \subset M$, there exists a constant C such that for all $f \in C_0^1(M)$ with $\operatorname{supp} f \Subset K$ we have

$$|F^\pm|_M^\infty \le C \|f\|_{C^1(M)}.$$

Remark 1.4.5. In the previous theorem, if f is in fact a *CR* function, then F is holomorphic on $U \setminus M$. We will employ the Plemelj formula in the proof of Lemma 2.4.3. For an idea of the proof, see, for example, [29].

Local extension

As we have seen, the problem of global extension of a *CR* function defined in the (connected) boundary of a compact domain of $\mathbb{C}^n$ is always solvable. The problem of local extension was first addressed by Hans Lewy [42] (but we remark that a previously published paper by Kneser [38] was ignored by the literature); the following result is one of the cornerstones of *CR* analysis.

Theorem 1.4.6. *Let S be a hypersurface in $\mathbb{C}^2$, of class C^2, defined by $\{\rho = 0\}$ (where $dp \neq 0$) and let $z^0 \in S$ be a point such that $\mathcal{L}_{z^0}(\rho)|_{H_{z^0}(S)} > 0$. Then there exists an open neighborhood U of z^0 in $\mathbb{C}^2$ such that every CR function $f : S \to \mathbb{C}$ of class C^1 extends by a function $\tilde{f}$ which is holomorphic in $U \cap \{\rho < 0\}$ and C^1 up to S.*

Lewy's result originated a vast amount of research, and was extended by Andreotti and Hill ([4, 5]) to $\mathbb{C}^n$ by cohomological methods. It turns out that a sufficient condition for one-sided extension in a neighborhood of $p \in S \subset \mathbb{C}^n$ is the presence of at least one non-vanishing eigenvalue in the Levi form $\mathcal{L}_p(S)$ (see [39]). If the Levi form has eigenvalues of both signs, then the bilateral extension (*i.e.* extension to a whole neighborhood of p in $\mathbb{C}^n$) of *CR* functions to holomorphic functions can be proved. The degenerate case when S is Levi flat was studied in [52, 53].

Semi-local extension

The local extension results for *CR* functions led naturally to the problem of "how far" these extensions could be defined. The developments in this directions were based on Bochner-Martinelli representation formula. In [44] the following result was proved:

Theorem 1.4.7. *Let Γ be a connected, compact hypersurface in $\mathbb{C}^n$, $n \geq 1$, of class C^1, with boundary $b\Gamma$. Assume that Γ is orientable and $b\Gamma$ satisfies the following conditions:*

(i) *$b\Gamma$ lies in the zero set M of a pluriharmonic function $\rho : \mathbb{C}^n \to \mathbb{R}$;*
(ii) *$\Gamma \setminus b\Gamma \subset \{z \in \mathbb{C}^n : \rho(z) > 0\}$;*
(iii) *$b\Gamma$ is the boundary of a bounded open subset A of M.*

Then, denoting by D the domain bounded by $b\Gamma \cup A$, every CR, locally Lipschitz function $f : \Gamma \setminus b\Gamma \to \mathbb{C}$ has a unique holomorphic extension F on D which is continuous on $D \cup (\Gamma \setminus b\Gamma)$.

We note that no (Levi) convexity condition is assumed on Γ, so that it is not a priori obvious even the extension of f to a one-sided neighborhood of Γ. The proof is achieved in a somewhat global way, by integration of the Bochner-Martinelli kernel on Γ and of a suitably chosen primitive of K_{BM} on $b\Gamma$.

The previous theorem was then generalized in [45], leading to the following result:

Theorem 1.4.8. *Let $D \Subset \mathbb{C}^n$ be a domain with boundary of class C^1, and let K be a compact subset of $\overline{D}$ such that*

- *K is $\mathcal{O}(\overline{D})$-convex;*
- *$bD \setminus K$ is connected.*

Then every continuous CR function f defined on $bD \setminus K$ extends to a unique $F \in \mathcal{O}(D \setminus K) \cap C^0(\overline{D} \setminus K)$.

In the recent years, these results have been further developed and extended to the case of Stein manifolds. Let X be a Stein manifold of dimension n ($n \geq 3$), and let $\Omega \subset X$ be a smooth, relatively compact, strictly pseudoconvex domain. Suppose that K is a compact subset of $\overline{\Omega}$ such that $\Omega \setminus K$ is pseudoconvex, and let $\omega = b\Omega \setminus K$. Consider the following conditions:

(A1) Every continuous CR function on ω extends to a (unique) continuous function on $E(\omega) = \overline{\Omega} \setminus K$ which is holomorphic in Ω.

(A2) For every smooth $(0, q)$-form f on ω ($1 \leq q \leq n - 3$) satisfying $\overline{\partial}_b f = 0$ there is a smooth $(0, q - 1)$-form u on ω such that $\overline{\partial}_b u = f$.

(A3) For every smooth $(0, n-2)$-form f on ω for which $\int_\omega f \wedge \varphi = 0$ for every smooth, $\overline{\partial}$-closed $(n, 1)$-form φ on $X \setminus K$ such that $\operatorname{supp} \varphi \cap \operatorname{supp} f$ is compact there is a smooth $(0, n-3)$-form u on ω satisfying $\overline{\partial}_b u = f$.

Then in [40] the following is proved:

Theorem 1.4.9. *If $\overline{\Omega \setminus K}$ is a Stein compact then*

$(A1)$, $(A2)$ *and* $(A3)$ *hold* $\iff H^{n,q}(K) = 0$ *for all* $1 \leq q \leq n - 1$.

In [27], geometric conditions are given for K and $\overline{\Omega} \setminus K$ to be Stein compacts and thus for (A1), (A2) and (A3) to hold, in the case that K is the union of all the connected components of $\Omega \setminus K$ except one, where $M \subset X$ is a Levi flat hypersurface. One of the main results of [27] is the following:

Theorem 1.4.10. *Assume that M is an orientable Levi-flat hypersurface of class C^3 in a complex manifold X. Let ρ be a strongly plurisubharmonic C^2 function in an open set $U \subset X$ such that the set $A = \{x \in U \cap M : \rho(x) \leq 0\}$ is compact. (Such A will be called a* compact strongly pseudoconvex set *in M.) If the Levi foliation of M is defined in a neighborhood of A by a nowhere vanishing* closed *one-form of class C^2 then A is a Stein compact.*

Afterwards, it is proved that in the same situation $\overline{\Omega} \setminus K$ is also a Stein compact, and thus

Corollary 1.4.11. *The condition in Theorem 1.4.10 implies that (A1), (A2), (A3) hold.*

Extension on unbounded domains

In another paper, Lupacciolu deals with the extension of *CR* functions defined on the boundary of not relatively compact domains Ω of $\mathbb{C}^n$. In [46], using Theorem 1.4.7, it is shown that - under a geometric hypothesis on Ω which constraints its behavior at infinity - the extension takes places:

Theorem 1.4.12. *Let $\Omega \subset \mathbb{C}^n$ be a (possibly unbounded) domain with boundary of class C^1. Suppose that*

$(\star)$ *there exists a polynomial $P \in \mathbb{C}[z_1, \ldots, z_n]$ such that $\Omega \subset \{z \in \mathbb{C}^n : |P(z)|^2 > (1 + \|z\|^2)^{\deg P}\}$.*

Then every continuous CR function f defined on $b\Omega$ extends to a function $F \in \mathcal{O}(\Omega) \cap C^0(\overline{\Omega})$.

We are going to use hypothesis $(\star)$ in Chapter 2. For now we only remark that the statement of Theorem 1.4.12 does not necessarily hold if condition $(\star)$ is removed. Here is an example.

Example 1.4.13. Let L be the complex hyperplane of $\mathbb{C}^n$ whose equation is $\{z_1 = 0\}$. Let $\Omega \subset \mathbb{C}^n$ be the domain defined by:

$$\Omega = \{\log |z_1|^2 + |z|^2 < 0\}.$$

$\Omega \supset L$ and it is immediate to check that $b\Omega$ is smooth and strongly pseudoconvex. Thus every CR function defined on $b\Omega$ extends to a holomorphic function defined on a neighborhood of $b\Omega$ in Ω. However, the function

$$f = \frac{1}{z_1}|_{b\Omega}$$

is trivially a CR function which is not extendable to the whole Ω (the holomorphic extension being uniquely determined).

Extension from minimal CR submanifolds

Let M be a CR manifold, and let $p \in M$. M is said to be *minimal* at p if there are no CR submanifolds of M through p of smaller dimension than that of M but with the same CR dimension (*i.e.* their holomorphic tangent space is actually the same as the one of M).

The following theorem by Tumanov [65] deals with the extension of CR functions from minimal submanifolds of $\mathbb{C}^n$. We say that M is W-extendable at $p \in M$ if there exists a wedge $\mathcal{W}$ with edge containing a neighborhood U of p in M (*i.e.* $\mathcal{W}$ is a product of U by some cone of $\mathbb{C}^n$) such that all CR functions on M holomorphically extend to $\mathcal{W}$.

Theorem 1.4.14. *A smooth, generic CR submanifold $M \subset \mathbb{C}^n$, minimal at $p \in M$, is W-extendable at p.*

We conclude by stating a fundamental result by Trépreau [64]. Let S be a real in $\mathbb{C}^n$ which separates, the space into two components Ω_+ and Ω_-, and let $z_0 \in S$. We say that Ω_i has *extension property* at z_0 if there exists a fundamental system of neighbourhoods U_n of z_0 such that all the holomorphic functions in $U_n \cap \Omega_j$ extend to U_n. Clearly, if S contains a germ of complex hypersurface through z_0, neither Ω_+ nor Ω_- has this property. The following result states that the converse also holds:

Theorem 1.4.15. *If S contains no germ of complex hypersurfaces at z_0 then at least one among Ω_+ and Ω_- has the local extension property at z_0.*

Chapter 2
Non-compact boundaries of complex subvarieties

2.1. Introduction

Let M be a smooth and oriented $(2m + 1)$-real submanifold of some n-complex manifold X. A natural question arises, whether M is the boundary of an $(m + 1)$-complex analytic subvariety of X. This problem, the so-called *boundary problem*, has been extensively treated over the past fifty years when M is compact and X is $\mathbb{C}^n$ or $\mathbb{CP}^n$.

The case when M is a compact, connected curve in $X = \mathbb{C}^n$ ($m = 0$), has been first solved by Wermer [66] in 1958. Later on, in 1975, Harvey and Lawson in [29] and [31] solved the boundary problem in $\mathbb{C}^n$ and then in $\mathbb{CP}^n \setminus \mathbb{CP}^r$, in terms of holomorphic chains, for any m. The boundary problem in $\mathbb{CP}^n$ was studied by Dolbeault and Henkin, in [20] for $m = 0$ and in [21] for any m. Moreover, in these two papers the boundary problem is dealt with also for closed submanifolds (with negligible singularities) contained in q-concave (*i.e.* union of $\mathbb{CP}^q$'s) open subsets of $\mathbb{CP}^n$. This allows M to be non compact. The results in [20] and [21] were extended by Dinh in [18].

A new approach to the problem in $\mathbb{CP}^n$ has been recently set forth by Harvey-Lawson [32–35]

The main theorem proved by Harvey and Lawson in [29] is that if $M \subset \mathbb{C}^n$ is compact and maximally complex then M is the boundary of a unique holomorphic chain of finite mass [29, Theorem 8.1]. If M is a connected strongly pseudoconvex submanifold, the solution is a complex subvariety with isolated *intrinsic* singularities, though there can be self-intersections. Moreover, if M is contained in the boundary $b\Omega$ of a strictly pseudoconvex domain Ω then M is the boundary of a complex analytic subvariety of Ω, with isolated singularities [30] (see also [28]). The converse of this result has also been proved, in the following setting: if $M \subset b\Omega$ is a $(2p - 1)$-submanifold and bounds a complex subvariety of Ω in the sense that there exist a proper holomorphic map $f : W \to \Omega$ ($\dim_{\mathbb{C}} W = p$) such that the closure of $f(W)$ is $f(W) \cup M$, then it is a maximally complex CR submanifold (see [25]).

The aim of this chapter is to generalize the previously mentioned result in [30] to a non compact, connected, closed and maximally complex submanifold M of the connected boundary $b\Omega$ of an unbounded weakly pseudoconvex domain $\Omega \subset \mathbb{C}^n$. The pseudoconvexity of Ω is needed both for the local result and to prove that the singularities are isolated. This result was obtained in a joint paper with A. Saracco [16].

Maximal complexity of M and the extension theorem for CR functions (see [42]) allow us to prove the following semi global result (see Corollary 2.3.3). Assume that $n \geq 3$, $m \geq 1$ and the Levi form $\mathcal{L}(b\Omega)$ of $b\Omega$ has at least $n - m$ positive eigenvalues at every point $p \in M$. Then

> there exist a tubular open neighborhood I of $b\Omega$ and a complex submanifold W_0 of $\overline{\Omega} \cap I$ with boundary, such that $bW_0 \cap b\Omega = M$, i.e. a complex manifold $W_0 \subset I \cap \Omega$ such that the closure $\overline{W}_0$ of W_0 in I is a smooth submanifold with boundary M.

A very simple example (see Example 2.3.4) shows that in general the semi global result fails to be true for $m = 0$.

In order to prove that W_0 extends to a complex analytic subvariety W of Ω with boundary M we first treat the case when Ω is convex and does not contain straight lines. This is the crucial step. For technical reasons we divide the proof in two cases: $m \geq 2$ and $m = 1$. We cut $\overline{\Omega}$ by a family of real hyperplanes H_λ which intersect M along smooth compact submanifolds. Then the natural foliation on each H_λ by complex hyperplanes induces on $M \cap H_\lambda$ a foliation by compact maximally complex $(2m - 1)$-real manifolds M'. Thus a natural way to proceed is to apply Harvey-Lawson's theorem to each slice M' and to show that the family $\{W'\}$ of the corresponding Harvey-Lawson solutions actually organizes in a complex analytic subvariety W, giving the desired extension (see Theorem 2.4.1). This is done following an idea of Zaitsev (see Lemma 2.4.3).

The same method of proof is used in the last section in order to treat the problem when Ω is more generally pseudoconvex. In this case, M is requested to fulfill an additional condition. Precisely,

($\star$) if $\overline{M}^\infty$ denotes the closure of $M \subset \mathbb{C}^n \subset \mathbb{CP}^n$ in $\mathbb{CP}^n$, then there exists an algebraic hypersurface V such that $V \cap \overline{M}^\infty = \emptyset$.

Equivalently

($\star'$) there exists a polynomial $P \in \mathbb{C}[z_1, \ldots, z_n]$ such that

$$M \subset \left\{ z \in \mathbb{C}^n : |P(z)|^2 > (1 + |z|^2)^{\deg P} \right\}.$$

A similar condition was first pointed out by Lupacciolu [46] in studying the extension problem for *CR* functions in unbounded domains. It allows us to build a nice family of hypersurfaces, which play the role of the hyperplanes in the convex case, and so to prove the main theorem of the paper:

Theorem 2.1.1. *Let Ω be an unbounded domain in $\mathbb{C}^n$ ($n \geq 3$) with smooth boundary $b\Omega$ and M be a maximally complex closed $(2m + 1)$-real submanifold ($m \geq 1$) of $b\Omega$. Assume that*

(i) *$b\Omega$ is weakly pseudoconvex and the Levi form $\mathcal{L}(b\Omega)$ has at least $n - m$ positive eigenvalues at every point of M;*
(ii) *M satisfies condition $(\star)$.*

Then there exists a unique $(m + 1)$-complex analytic subvariety W of Ω, such that $bW = M$. Moreover the singular locus of W is discrete and the closure of W in $\overline{\Omega} \setminus \text{Sing}(W)$ is a smooth submanifold with boundary M.

We do not deal with the 1-dimensional case. There are two different kinds of difficulties. First of all, a semi global strip as in Corollary 2.3.3 may not exist (see Example 2.3.4). Secondly, even if it does exist, it could be non extendable to the whole Ω (see Example 2.4.7) and it is not clear at all how it is possible to generalize the *moments condition* (see [29]).

Another similar approach can be followed to treat the *semi-local* boundary problem, *i.e.* given an open subset U of the boundary of Ω, find an open subset $\Omega' \subset \Omega$ such that, for any maximally complex submanifold $M \subset U$, there exists a complex subvariety W of Ω' whose boundary is M. We deal with this problem in Chapter 3.

2.2. Definitions and notations

We recall here the notions of *CR* geometry that will be used in the through-out the chapter, along with the notations we will employ. We refer to Chapter 1 for a more complete description.

Let $N \subset \mathbb{C}^n$ be a smooth connected real submanifold, and let $p \in N$. We will denote by $T_p(N)$ the tangent space of N at the point p, and by $H_p(N)$ the holomorphic tangent space of N at the point p.

If N is a *CR* submanifold and $\dim_{\mathbb{C}} H_p(N)$ is the greatest possible, *i.e.* $\dim_{\mathbb{C}} H_p(N) = k$ for every p, N is said to be *maximally complex*.

We say that a C^∞ function $f : N \to \mathbb{C}$ is a *CR function* if for a C^∞ extension (and hence for any) $\tilde{f} : U \to \mathbb{C}$ (U being a neighborhood of N) we have

$$\left(\overline{\partial}\tilde{f}\right)\big|_{H(N)} = 0. \qquad (2.2.1)$$

In particular the restriction of a holomorphic function to a CR submanifold is a CR function. It is immediately seen that f is CR if and only if

$$df \wedge (dz_1 \wedge \ldots \wedge dz_n)|_N = 0. \tag{2.2.2}$$

Similarly N is maximally complex if and only if

$$(dz_{j_1} \wedge \ldots \wedge dz_{j_{k+1}})|_N = 0,$$

for any $(j_1, \ldots, j_{k+1}) \in \{1, \ldots, n\}^{k+1}$.

Finally we observe that the boundary M of a complex submanifold W with $\dim_{\mathbb{C}} W > 1$ is maximally complex. Indeed, for any $p \in bW = M$, $T_p(bW)$ is a real hyperplane of $T_p(W) = H_p(W)$ and so is $J(T_p(bW))$. Hence $H_p(bW) = T_p(bW) \cap J(T_p(bW))$ is of real codimension 2 in $H_p(W)$.

If $\dim_{\mathbb{C}} W = 1$ and bW is compact then for any holomorphic $(1, 0)$-form ω we have

$$\int_M \omega = \iint_W d\omega = \iint_W \partial\omega = 0,$$

since $\partial\omega|_W \equiv 0$. This condition for M is called *moments condition* (see [29]).

2.3. The local and semi global results

The aim of this section is to prove the local result. Given a smooth real hypersurface S in $\mathbb{C}^n$, we denote by $\mathcal{L}_p(S)$ the Levi form of S at the point p. Let 0 be a point of M. We have the following inclusions of tangent spaces:

$$\mathbb{C}^n \supset T_0(S) \supset H_0(S) \supset H_0(M);$$

$$T_0(S) \supset T_0(M) \supset H_0(M).$$

Lemma 2.3.1. *Let M be a maximally complex submanifold of a smooth real hypersurface S, $\dim_{\mathbb{R}} M = 2m + 1$, $m \geq 1$, $0 \in M$. Suppose that $\mathcal{L}_0(S)$ has at least $n - m$ eigenvalues of the same sign. Then*

$$H_0(S) \not\supset T_0(M).$$

Proof. By hypothesis $\mathcal{L}_0(S)$ is strictly positive definite on a non-zero subspace P of $H_0(M)$. Let $X_0 \in P^{\mathbb{C}} \cap H_0^{1,0}$, and let $\widetilde{X}$ be a local section of $H(S)$ such that $\widetilde{X}(0) = X_0$ and $\widetilde{X}(p) \in H(M)$ when $p \in M$. Then (the projection on $T_0(S)$ of) $[\widetilde{X}, \overline{\widetilde{X}}](0)$ does not belong to $H_0(S)$, but it does to $T_0(M)$. $\square$

Lemma 2.3.2. *Under the hypothesis of Lemma 2.3.1, assume that S is the boundary of an unbounded domain $\Omega \subset \mathbb{C}^n$, $0 \in M$ and that the Levi form of S has at least $n - m$ positive eigenvalues. Then*

(i) *there exists an open neighborhood U of 0 and an $(m + 1)$-complex submanifold $W_0 \subset U$ with boundary, such that $bW_0 = M \cap U$;*

(ii) $W_0 \subset \Omega \cap U$.

Proof. To obtain $\mathcal{L}_0(M)(\zeta_0, \overline{\zeta}_0)$ we choose a smooth local section ζ of $H(M)$ around 0, such that $\zeta(0) = \zeta_0$, and compute the projection of the bracket $[\zeta, \overline{\zeta}](0)$ on the real part of $T_0(M)$. By hypothesis, the intersection of the space where $\mathcal{L}_0(S)$ is positive with $H_0(M)$ is non empty; take η_0 in this intersection. Then $\mathcal{L}_0(M)(\eta_0, \overline{\eta}_0) \neq 0$. Suppose, by contradiction, that the bracket $[\eta, \overline{\eta}](0)$ lies in $H_0(M)$, *i.e.* its projection on the real part of the tangent of M is zero. Then, if $\widetilde{\eta}$ is a local smooth extension of the field η to S, we have $[\widetilde{\eta}, \overline{\widetilde{\eta}}](0) = [\eta, \overline{\eta}](0) \in H_0(M)$. Since $H_0(M) \subset H_0(S)$, this would mean that the Levi form of S in 0 is zero in η_0. Now, we project (generically) M over a $\mathbb{C}^{m+1}$ in such a way that the projection π is a local embedding near 0: since the restriction of π to M is a CR map, and since the Levi form of M has - by the arguments stated above - at least one positive eigenvalue, it follows that the Levi form of $\pi(M)$ has at least one positive eigenvalue. Thus, in order to obtain W_0, it is sufficient to apply the Lewy extension theorem [42] (or rather the version in $\mathbb{C}^n$, see [4], [5]) to the CR function $\pi^{-1}|_M$.

As for the second statement, we observe that the projection by π of the normal vector of S pointing towards Ω lies into the domain of $\mathbb{C}^{m+1}$ where the above extension W_0 is defined. Indeed, the Lewy extension result gives a holomorphic function in the connected component of (a neighborhood of 0 in) $\mathbb{C}^n \setminus \pi(M)$ for which $\mathcal{L}_0(\pi(M))$ has a positive eigenvalue when $\pi(M)$ is oriented as the boundary of this component. This is precisely the component towards which the projection of the normal vector of S points when the orientations of S and M are chosen accordingly. This fact, combined with Lemma 2.3.1 (which states that any extension of M must be transverse to S) implies that locally $W_0 \subset \Omega \cap U$. $\qquad\square$

Corollary 2.3.3 (Semi global existence of W). *Under the same hypothesis of Lemma 2.3.2, there exist an open tubular neighborhood I of S in $\overline{\Omega}$ and an $(m + 1)$-complex submanifold W_0 of $\overline{\Omega} \cap I$, with boundary, such that $S \cap bW_0 = M$.*

Proof. By Lemma 2.3.2, for each point $p \in M$, there exist a neighborhood U_p of p and a complex manifold $W_p \subset \overline{\Omega} \cap U_p$ bounded by M. We cover M with countable many such open sets U_i, and consider the union $W_0 = \cup_i W_i$. W_0 is contained in the union of the U_i's, hence we may restrict it to a tubular neighborhood I_M of M. It is easy to extend I_M to a tubular neighborhood I of S.

The fact that $W_i|_{U_{ij}} = W_j|_{U_{ij}}$ if $U_i \cap U_j = U_{ij} \neq \emptyset$ immediately follows from the construction made in Lemma 2.3.2, in view of the uniqueness of the holomorphic extension of CR functions. $\qquad\square$

Example 2.3.4. Corollary 2.3.3 could be restated by saying that if a submanifold $M \subset S$ ($\dim_{\mathbb{R}} M \geq 3$) is locally extendable at each point as a complex manifold, then (one side of) the extension lies in Ω. This is no longer true, in general, for curves, as shown in $\mathbb{C}^n_{(z_1,\dots,z_{n-1},w)}$, $z_k = x_k + iy_k$, $w = u + iv$, by the following case:

$$S = \left\{ v = u^2 + \sum_k |z_k|^2 \right\}, \quad \Omega = \left\{ v > u^2 + \sum_k |z_k|^2 \right\},$$

$$M = \left\{ y_1 = 0, \ v = x_1^2, \ u = 0, \ z_2 = \cdots = z_{n-1} = 0 \right\}$$

and

$$W = \left\{ w = iz_1^2, \ z_2 = \cdots = z_{n-1} = 0 \right\};$$

we have that $S \cap W = M$ and $W \subset \mathbb{C}^n \setminus \Omega$.

Remark 2.3.5. Suppose that S is strongly pseudoconvex and choose, in $\mathbb{C}^n_{(z_1,\dots,z_n)}$, a local strongly plurisubharmonic equation ρ for S: $S = \{\rho = 0\}$. Consider the curve

$$\gamma = \{z_j = \gamma_j(t), \ j = 1,\dots,n, \ t \in (-\varepsilon, \varepsilon)\} \subset S.$$

Assume that γ is real analytic, so that locally there exists a complex curve $\tilde{\gamma}$ such that $\tilde{\gamma} \supset \gamma$. Then one side of $\tilde{\gamma}$ lies in Ω if and only if

$$\sum_j \operatorname{Re} \frac{\partial \rho}{\partial z_j} \frac{\partial \gamma_j}{\partial t} \neq 0. \tag{2.3.1}$$

Observe that condition (2.3.1), which depends only on γ (when S is given), is not satisfied in Example 2.3.4. The sufficiency of (2.3.1) for extension is true for *any* real hypersurface S: indeed, from a geometric point of view, the condition is equivalent to the transversality of $T(\tilde{\gamma})$ and $H(S)$ (and hence $T(S)$). Pseudoconvexity is required to establish the necessity.

2.4. The global result

In order to make the proof more transparent we first treat the case when Ω is an unbounded convex domain with smooth boundary $b\Omega$. In the next section we will prove the main theorem in all its generality.

Theorem 2.4.1. *Let M be a maximally complex (connected) $(2m + 1)$-real submanifold $(m \geq 1)$ of $b\Omega$. Assume that Ω does not contain straight lines and $b\Omega = S$ satisfies the conditions of Lemma 2.3.1. Then there exists an $(m + 1)$-complex subvariety W of Ω, with isolated singularities, such that $bW = M$.*

We observe that under the hypothesis of Theorem 2.4.1, there exists a complex strip in a tubular neighborhood with boundary M (see Corollary 2.3.3). Moreover, since Ω does not contain straight lines, we can approximate uniformly from both sides $b\Omega$ by strictly convex domains, see [50]. It follows that we can find a real hyperplane L such that, for any translation L' of L, $L' \cap \overline{\Omega}$ is a compact set. Now choose an exhaustive sequence L_k of such hyperplanes, and denote by Ω_k the bounded connected component of $\Omega \setminus L_k$. Then, approximating from inside, we can choose a strictly convex open subset $\Omega'_k \subset \Omega$ such that $b\Omega'_k \cap \Omega_k \subset I$, where I is the tubular neighborhood of Corollary 2.3.3. Then, it is easily seen that we are in the situation of the following

Proposition 2.4.2. *Let $D \Subset B \Subset \mathbb{C}^n$ $(n \geq 4)$ be two strictly convex domains. Let $D_+ = D \cap \{\mathrm{Re}\, z_n > 0\}$, $B_+ = B \cap \{\mathrm{Re}\, z_n > 0\}$. Then every $(m + 1)$-complex subvariety $(m \geq 2)$ with isolated singularities, $A \subset B_+ \setminus \overline{D}_+ = C_+$, is the restriction of a complex subvariety $\widetilde{A}$ of B_+ with isolated singularities.*

We treat the cases $m \geq 2$ and $m = 1$ separately. Indeed all the main ideas of the proof are present in the case $m \geq 2$. The case $m = 1$ simply adds technical difficulties.

2.4.1. *M* is of dimension at least 5: $m \geq 2$

Before proving Proposition 2.4.2, we need some preliminary considerations and to prove two Lemmas.

Let φ be a strictly convex function[1] defined in a neighborhood of B such that $B = \{\varphi < 0\}$. Fixing $\varepsilon > 0$ small enough, $B' = \{\varphi < -\varepsilon\}$ is a strictly convex domain of B whose boundary H intersects A in a smooth maximally complex submanifold N. A natural way to proceed is to slice

---

[1] In the general case φ will be a strongly plurisubharmonic function.

N with complex hyperplanes, in order to apply Harvey-Lawson's theorem. Each slice of B' is strictly convex, hence strongly pseudoconvex, and so the holomorphic chain that we obtain is contained in B'. Thus the set made up by collecting the chains is contained in B'. Analyticity of this set is the hard part of the proof.

Because of Sard's Lemma, for all $z \in D_+$, there exist a vector v arbitrarily close to $\partial/\partial z_n$, and $k \in \mathbb{C}$ such that z belongs to the hyperplane v_k defined as $v^\perp + k$ and $A_k = v_k \cap N$ is transversal and compact, and thus smooth.

In a neighborhood of each fixed $z_0 \in D_+$, the same vector v realizes the transversality condition. Hence we should now fix our attention to a neighborhood of the form $\widehat{U} = \bigcup_{k \in U} v_k \cap B_+$, where v_{k_0} is the affine hyperplane containing z_0 and $U \subset \mathbb{C}$ a neighborhood of k_0.

Let $\pi : \widehat{U} \to \mathbb{C}^m$ be a generic projection: we use (w', w) as holomorphic coordinates on $v_{k_0} = \mathbb{C}^m \times \mathbb{C}^{n-m-1}$ (and also for k near to k_0). Let $V_k = \mathbb{C}^m \setminus \pi(A_k)$, and $V = \cap_k V_k$.

Since A_{k_0} has a local extension (given by $v_{k_0} \cap A$), it is maximally complex and so, by Harvey-Lawson's theorem, there is a holomorphic chain $\widetilde{A}_{k_0}$ with $b\widetilde{A}_{k_0} = A_{k_0}$, which extends holomorphically A_{k_0}.

Our goal is to show that $\widetilde{A}_U = \cup_k \widetilde{A}_k$ is analytic in $\pi^{-1}(V)$. From this, it will follow that $\widetilde{A}_U$ is an analytic subvariety of $\widehat{U}$, π being a generic projection.

Following an idea of Zaitsev, for $k \in U$, $w' \in \mathbb{C}^m \setminus \pi(A_k)$ and $\alpha \in \mathbb{N}^{n-m-1}$, we define

$$I^\alpha(w', k) = \int_{(\eta', \eta) \in A_k} \eta^\alpha \omega_{BM}(\eta' - w'),$$

ω_{BM} being the Bochner-Martinelli kernel.

(Note that our notation here varies slightly with respect to the one employed in the first chapter: $\omega_{BM}(\eta' - w') = K_{BM}(\eta', w')$).

Lemma 2.4.3. *Let $F(w', k)$ be the multiple-valued function which represents $\widetilde{A}_k$ on $\mathbb{C}^m \setminus \pi(A_k)$. Then, if we denote by $P^\alpha(F(w', k))$ the sum of the αth powers of the values of $F(w', k)$, the following holds:*

$$P^\alpha(F(w', k)) = I^\alpha(w', k).$$

In particular, $F(w', k)$ is finite.

Proof. Let V_0 be the unbounded component of V_k (where, of course, $P^\alpha(F(w', k)) = 0$). It is easy to show, following [29], that on V_0 also $I^\alpha(F(w', k)) = 0$: in fact, if w' is far enough from $\pi(A_k)$, then $\beta =$

$\eta^{\alpha}\omega_{BM}(\eta' - w')$ is a regular $(m, m-1)$-form on some Stein neighborhood O of A_k. So, since in O there exists γ such that $\overline{\partial}\gamma = \beta$, we may write in the language of currents

$$[A_k](\beta) = [A_k]_{m,m-1}(\overline{\partial}\gamma) = \overline{\partial}[A_k]_{m,m-1}(\gamma) = 0.$$

In fact, since A_k is maximally complex, $[A_k] = [A_k]_{m,m-1} + [A_k]_{m-1,m}$ and $\overline{\partial}[A_k]_{m,m-1} = 0$, see [29]. Moreover, since $[A_k](\beta)$ is analytic in the variable w', $[A_k](\beta) = 0$ for all $w' \in V_0$.

To conclude our proof, we just need to show that the "jumps" of the functions $P^{\alpha}(F(w', k))$ and $I^{\alpha}(w', k)$ across the regular part of the common boundary of two components of V_k are the same.

So, let $z' \in \pi(A_k)$ be a regular point in the common boundary of V_1 and V_2. Locally in a neighborhood of z', we can write $\widetilde{A}_k$ as a finite union of graphs of holomorphic functions, whose boundaries A_k^i are either in A_k or empty. In the first case, the A_k^i are CR graphs over $\pi(A_k)$ in the neighborhood of z'. We may thus consider the jump j_i of $P^{\alpha}(F(w', k))$ due to a single function. We remark that the jump for a function f is $j_i = f(z')^{\alpha}$. The total jump will be the sum of them.

To deal with the jump of $I^{\alpha}(w', k)$ across z', we split the integration set in the sets A_k^i (thus obtaining the integrals I_i^{α}) and $A_k \setminus \cup_i A_k^i$ (I_0^{α}). Thanks to Plemelj's formulas (see [29]) the jumps of I_i^{α} are precisely j_i. Moreover, since the form $\eta^{\alpha}\omega_{BM}(\eta' - z')$ is C^{∞} in a neighborhood of $A_k \setminus \cup_i A_k^i$, the jump of I_0^{α} is 0. So $P^{\alpha}(F(w', k)) = I^{\alpha}(w', k)$.   □

Remark 2.4.4. Lemma 2.4.3 implies, in particular, that the functions $P^{\alpha}(F(w', k))$ are continuous in k. Indeed, they are represented as integrals of a fixed form over submanifolds A_k which vary continuously with the parameter k.

The functions $P^{\alpha}(F(w', k))$ and the holomorphic chain $\widetilde{A}_{k_0}$ uniquely determine each other and so, proving that the union over k of the $\widetilde{A}_k$ is an analytic set is equivalent to proving that the functions $P^{\alpha}(F(w', k))$ are holomorphic in the variable $k \in U \subset \mathbb{C}$.

Lemma 2.4.5. $P^{\alpha}(F(w', k))$ is holomorphic in the variable $k \in U \subset \mathbb{C}$, for each $\alpha \in \mathbb{N}^{n-m-1}$.

Proof. Let us fix a point $(w', \underline{k})$ such that $w' \notin A_{\underline{k}}$ (this condition remains true for $k \in B_{\epsilon}(\underline{k})$). Consider as domain of $P^{\alpha}(F)$ the set $\{w'\} \times B_{\epsilon}(\underline{k})$. In view of Morera's theorem, we need to prove that for any simple curve $\gamma \subset B_{\epsilon}(\underline{k})$,

$$\int_{\gamma} P^{\alpha}(F(w', k))dk = 0.$$

Let $\Gamma \subset B_\epsilon(\underline{k})$ be an open set such that $b\Gamma = \gamma$. By $\gamma * A_k$ ($\Gamma * A_k$) we mean the union of A_k along γ (along Γ). Note that these sets are submanifolds of N ($\Gamma * A_k$ is an open subset) and $b(\Gamma * A_k) = \gamma * A_k$. By Lemma 2.4.3 and Stoke's theorem

$$
\begin{aligned}
\int_\gamma P^\alpha(F(w',k))dk &= \int_\gamma I^\alpha(w',k)dk \\
&= \int_\gamma \left(\int_{(\eta',\eta) \in A_k} \eta^\alpha \omega_{BM}(\eta' - w') \right) dk \\
&= \iint_{\gamma * A_k} \eta^\alpha \omega_{BM}(\eta' - w') \wedge dk \\
&= \iint_{\Gamma * A_k} d\left(\eta^\alpha \omega_{BM}(\eta' - w') \wedge dk \right) \\
&= \iint_{\Gamma * A_k} d\eta^\alpha \wedge \omega_{BM}(\eta' - w') \wedge dk \\
&= 0.
\end{aligned}
$$

The last equality follows from the fact that since η^α is holomorphic, only holomorphic differentials appear in $d\eta^\alpha$. Consequently, since all the holomorphic differentials supported by $\Gamma * A_k$ already appear in $\omega_{BM}(\eta' - w') \wedge dk$, the integral is zero. $\square$

Now, we are in position to prove Proposition 2.4.2.

Proof of Proposition 2.4.2 ($m \geq 2$). Up to this point we have extended the complex manifold A to an analytic set

$$
\widetilde{A}_U \doteq A \cup \bigcup_{k \in U} \widetilde{A}_k \subset V_U \doteq C_+ \cup \bigcup_{k \in U} (v_k \cap B_+).
$$

The open sets V_U are an open covering of B_+.

Moreover the open sets $\omega_U \doteq \bigcup_{k \in U}(v_k \cap B_+)$ form an open covering of each compact set $K_\delta \doteq \overline{B}' \cap \{\mathrm{Re}\, z_n \geq \delta\}$. Hence there exist $\omega_1, \ldots, \omega_l$ which cover K_δ and such that $\omega_i \cap \omega_{i+1} \cap C_+ \neq \emptyset$, for $i = 1, \ldots, l - 1$ and therefore there exists a countable open cover $\{\omega_i\}_{i \in \mathbb{N}}$ of $\overline{B}' \cap B_+$ such that, for all $i \in \mathbb{N}$, $\omega_i \cap \omega_{i+1} \cap C_+ \neq \emptyset$.

So we may extend A to $C_+ \cup \omega_1$ by proceeding as above.

Suppose now that we have extended A to $C^i \doteq C_+ \cup \bigcup_{j=1}^i \omega_j$ with an analytic set A_i. On the non-empty intersection $C^i \cap \omega_{i+1} \cap C_+$ A_i and the extension $\widetilde{A}_{i+1}$ of A to $C_+ \cup \omega_{i+1}$ coincide (as they both coincide with

A), hence by analicity they coincide everywhere. Consequently we may extend A to C^{i+1} by $A_{i+1} \doteq A_i \cup \widetilde{A}_{i+1}$. It follows that

$$\widetilde{A} \doteq A \cup \bigcup_{j \in \mathbb{N}} A_j,$$

is the desired extension of A to B_+.

In order to conclude the proof we have to show that $\widetilde{A}$ has isolated singularities. Let $\mathsf{Sing}\,(\widetilde{A}) \subset B'_+$ be the singular locus of $\widetilde{A}$.

Recall that φ is a strictly convex defining function for B. Let us consider the family

$$(\phi_\lambda = \lambda \varphi + (1 - \lambda)\mathsf{Re}\, z_n)_{\lambda \in [0,1]}$$

of strictly convex functions. For λ near to 1, $\{\phi_\lambda = 0\}$ does not intersect the singular locus $\mathsf{Sing}\,(\widetilde{A})$. Let $\bar{\lambda}$ be the biggest value of λ for which

$$\{\phi_\lambda = 0\} \cap \mathsf{Sing}\,(\widetilde{A}) \neq \emptyset.$$

Then

$$\left\{\phi_{\bar{\lambda}} < 0\right\} \cap B_+ \subset B_+$$

is a Stein domain in whose closure the analytic set $\mathsf{Sing}\,(\widetilde{A})$ is contained, touching the boundary in a point of strict convexity. So, by Kontinuitätsatz,

$$\{\phi_{\bar{\lambda}} = 0\} \cap \mathsf{Sing}\,(\widetilde{A})$$

is a set of isolated points in $\mathsf{Sing}\,(\widetilde{A})$. By repeating this argument, we conclude that $\mathsf{Sing}\,(\widetilde{A})$ is made up by isolated points.   □

Proof of Theorem 2.4.1 ($m \geq 2$). Thanks to Corollary 2.3.3, we have a regular submanifold W_1 of a tubular neighborhood I, with boundary M.

Suppose $0 \in M$. The real hyperplanes $H_k \doteq T_0(S) + k$, $k \in \mathbb{R}$, intesect S in a compact set. If the intersection is non-empty, $\bar{\Omega}$ is divided in two sets. Let Ω_k be the compact one. We can choose a sequence H_{k_n} such that Ω_{k_n} is an exaustive sequence for $\bar{\Omega}$.

We apply Proposition 2.4.2 with $B_+ = \Omega_{k_n}$, $C_+ = I \cap \Omega_{k_n}$, and $A = W_1 \cap \Omega_{k_n}$, to obtain an extension of W_1 in Ω_{k_n}. Since, by the identity principle, two such extensions coincide in $\Omega_{k_{\min\{n,m\}}}$, their union is the desired submanifold W.   □

2.4.2. M is of dimension 3 $(m = 1)$

We prove now the statement of Proposition 2.4.2 for $m = 1$.

Our first step is to show that when we slice transversally N with complex hyperplanes, we obtain 1-real submanifolds which satisfy the moments condition.

Again, we fix our attention to a neighborhood of the form

$$\widehat{U} \doteq \bigcup_{k \in U} v_k \cap B_+.$$

In $\widehat{U}$, with coordinates $w_1, \ldots, w_{n-1}, k$, we choose an arbitrary holomorphic $(1, 0)$-form which is constant with respect to k.

Lemma 2.4.6. *The function*

$$\Phi_\omega(k) = \int_{A_k} \omega$$

is holomorphic in U.

Proof. We again use Morera's theorem. We need to prove that for any simple curve $\gamma \subset U$, $\gamma = b\Gamma$,

$$\int_\gamma \Phi_\omega(k)dk = 0.$$

By virtue of Stoke's theorem, we have

$$
\begin{aligned}
\int_\gamma \Phi_\omega(k)dk &= \int_\gamma \left(\int_{A_k} \omega \right) dk \\
&= \iint_{\gamma * A_k} \omega \wedge dk \\
&= \iint_{\Gamma * A_k} d(\omega \wedge dk) \\
&= \iint_{\Gamma * A_k} \partial\omega \wedge dk \\
&= 0.
\end{aligned}
$$

The last equality is due to the fact that $\Gamma * A_k \subset N$ is maximally complex and thus supports only $(2, 1)$ and $(1, 2)$-forms, while $\partial\omega \wedge dk$ is a $(3, 0)$-form. $\qquad\square$

Now we can prove Proposition 2.4.2 and Theorem 2.4.1 when $m = 1$.

We can find a countable covering of B_+ made of open subsets $\omega_i = \widehat{U}_i \cap B_+$ in such a way that:

1. $\omega_0 \subset C_+$;
2. if

$$B_l = \bigcup_{i=1}^{l} \omega_i,$$

then $\omega_{l+1} \cap B_l \supset v_{l+1} \cap B_+$, where v_{l+1} is a complex hyperplane in $\widehat{U}_{l+1}$.

Suppose that we have already found $\widetilde{A}_l$ extending A on B_l (observe that in $B_0 = \omega_0$, $\widetilde{A}_0 = A$). To conclude the proof we have to find $\widetilde{A}_{l+1}$ extending A on B_{l+1}.

Each slice of N in B_l is maximally complex, and so are $v_{l+1} \cap N$ and $v_\epsilon \cap N$, for $v_\epsilon \subset \omega_{l+1}$ sufficiently near to v_{l+1} (because they are in B_l as well).

Now we use Lemma 2.4.6 with $\widehat{U} = \widehat{U}_{l+1}$. What we have just observed implies that, for all holomorphic $(1, 0)$-form η, $\Phi_\eta(k)$ vanishes on an open subset of U and so is identically zero on U. This implies that all slices in ω_{l+1} are maximally complex. Again we may apply Harvey-Lawson's theorem slice by slice and conclude by the methods of Proposition 2.4.2.

2.4.3. *M* is of dimension 1: $m = 0$

We have already observed that if M is one-dimensional the local extension inside Ω may not exist (see Example 2.3.4). Even though there is a local strip in which we have an extension, the methods used to prove Proposition 2.4.2 do not work, since the transversal slices M are either empty or isolated points. Indeed, as the following example shows, that extension result does not hold for $m = 0$.

Example 2.4.7. Using the notations of Proposition 2.4.2, consider in $\mathbb{C}^2$ the balls

$$B = \left\{ |z_1|^2 + |z_2|^2 < c \right\}, \quad D = \left\{ |z_1|^2 + |z_2|^2 < \varepsilon \right\}, \quad c > \varepsilon > 2.$$

Let A be the connected irreducible complex curve

$$\{(z_1, z_2) \in B_+ \ : \ z_1 z_2 = 1\}$$

and set $A_C = A \cap C_+$.

A_C has two connected components. Indeed, a point of A (of A_C) can be written as $z_1 = \rho e^{i\theta}$, $z_2 = \frac{1}{\rho} e^{-i\theta}$, with $\rho \in \mathbb{R}^+$ and $\theta \in \left(-\frac{\pi}{2}, \frac{\pi}{2}\right)$. Hence, points in A_C satisfy

$$2 < \varepsilon < \rho^2 + \frac{1}{\rho^2} < c \;\Rightarrow\; 2 < \sqrt{\varepsilon + 2} < \rho + \frac{1}{\rho} < \sqrt{c + 2}.$$

Since $f(\rho) = \rho + 1/\rho$ is monotone decreasing up to $\rho = 1$ (where $f(1) = 2$), and then monotone increasing, there exist a and b such that the inequalities are satisfied when $a < \rho < b < 1$, or when $1 < 1/b < \rho < 1/a$. A_C is thus the union of the two disjoint open sets

$$A_1 = \left\{ \left(\rho e^{i\theta}, \tfrac{1}{\rho} e^{-i\theta}\right) \in \mathbb{C}^2 \;\middle|\; a < \rho < b, \; -\tfrac{\pi}{2} < \theta < \tfrac{\pi}{2} \right\};$$

$$A_2 = \left\{ \left(\rho e^{i\theta}, \tfrac{1}{\rho} e^{-i\theta}\right) \in \mathbb{C}^2 \;\middle|\; a < \tfrac{1}{\rho} < b, \; -\tfrac{\pi}{2} < \theta < \tfrac{\pi}{2} \right\}.$$

It follows that A_1 is not extendable to B as an analytic subset. Indeed, since A is irreducible, such an extension of A_1 should contain A_2.

2.5. Extension to pseudoconvex domains

We may now prove the following

Theorem 2.5.1. *Let Ω be an unbounded domain in $\mathbb{C}^n$ ($n \geq 3$) with smooth boundary $b\Omega$ and M a maximally complex closed $(2m + 1)$-real submanifold ($m \geq 1$) of $b\Omega$. Assume that*

(i) *$b\Omega$ is weakly pseudoconvex and the Levi form $\mathcal{L}(b\Omega)$ has at least $n - m$ positive eigenvalues at every point of M;*
(ii) *M satisfies condition $(\star)$.*

Then there exists a unique $(m + 1)$-complex analytic subvariety W of Ω, such that $bW = M$. Moreover, the singular locus of W is discrete and the closure of W in $\overline{\Omega} \setminus \mathrm{Sing}\, W$ is a smooth submanifold with boundary M.

Proof. Assume, for the moment, that condition $(\star)$ is replaced by the stronger one

$$\overline{\Omega}^\infty \cap \Sigma_0 = \emptyset \text{ where } \overline{\Omega}^\infty \text{ denotes the projective closure of } \Omega.$$

In order to conclude the proof (using the methods of the previous section) the only thing to show is that, up to a holomorphic change of coordinates and a holomorphic embedding $V : \mathbb{C}^n \to \mathbb{C}^N$, we can choose a sequence of real hyperplanes $H_k \subset \mathbb{C}^N$, $k \in \mathbb{N}$, which is exhaustive in the following sense:

1. $H_k \cap V(S)$ is compact, for all $k \in \mathbb{N}$;
2. one of the two halfspaces in which H_k divides $\mathbb{C}^N$, say H_k^+, intersects $V(\Omega)$ in a relatively compact set;
3. $\bigcup_k (H_k^+ \cap V(\Omega)) = V(\Omega)$.

The arguments of Proposition 2.4.2, indeed - excluded the proof that the singularities are isolated - depend only on the fact that we can cut M by complex hyperplanes, obtaining compact maximally complex submanifolds. Once we have found $W' \subset V(\mathbb{C}^n)$ (W' is in fact contained in $V(\mathbb{C}^n)$ by analytic continuation, since it has to coincide with the strip in a neighborhood of M), we set $W = V^{-1}(W')$.

Observe that the hypersurfaces $V^{-1}(H_k)$ form an exhaustive sequence for Ω; let Ω_k be correspondent sequence of relatively compact subsets. Since Ω is a domain of holomorphy, for each k we can choose a strongly pseudoconvex open subset $\Omega_k' \subset \Omega$ such that $b\Omega_k' \cap \Omega_k \subset I$, where I is the tubular neighborhood found in Corollary 2.3.3. So, in each Ω_k we can suppose that we deal with a strongly pseudoconvex open set, and thus the proof of the fact that the singularities are isolated is the same as in Proposition 2.4.2.

Following [46] we divide the proof in two steps.

Step 1. P linear. We consider $\overline{\Omega} \subset \mathbb{CP}^n = \mathbb{C}^n \cup \mathbb{CP}_\infty^{n-1}$, which is disjoint from $\Sigma_0 = \{P = 0\}$. So we can consider new homogeneous coordinates $[z_0 : \ldots : z_n]$ of $\mathbb{CP}^n$ in such a way that Σ_0 is the hyperplane $\mathbb{CP}^{n-1}$ at infinity, $\mathbb{CP}^{n-1} = \{z_0 = 0\}$. Now Ω is a relatively compact open set of $(\mathbb{C}^n)' = \mathbb{CP}^n \setminus \Sigma_0$, and $H_\infty = \mathbb{CP}_\infty^{n-1} \cap (\mathbb{C}^n)'$ is a complex hyperplane containing the boundary of S. Let $H_\infty^{\mathbb{R}} \supset H_\infty$ be a real hyperplane. The intersection between S and a translated of $H_\infty^{\mathbb{R}}$ is either empty or compact. For all $z \in \Omega$, there exist a real hyperplane $H_\infty^{\mathbb{R}} \not\ni z$, intersecting Ω, and a small translated H_{ε_z} such that $z \in H_{\varepsilon_z}^+$. Since $\Omega = \bigcup_z (H_{\varepsilon_z}^+ \cap \Omega)$, and Ω is a countable union of compact sets, we may choose an exhaustive sequence H_k.

Step 2. P generic. We use the Veronese map v to embed $\mathbb{CP}^n$ in a suitable $\mathbb{CP}^N$ in such a way that $v(\Sigma_0) = L_0 \cap v(\mathbb{CP}^n)$, where L_0 is a linear subspace. The *Veronese map* v is defined as follows: let d be the degree of P, and let

$$N = \binom{n+d}{d} - 1.$$

Then $v : \mathbb{CP}^n \to \mathbb{CP}^N$ is defined by

$$v(z) = v[z_0 : \ldots : z_n] = [\ldots : w_I : \ldots]_{|I|=d},$$

where $w_I = z^I$. If $P = \sum_{|I|=d} \alpha_I z^I$, then $v(\Sigma_0) = L_0 \cap v(\mathbb{CP}^n)$, where

$$L_0 = \left\{ \sum_{|I|=d} \alpha_I w_I = 0 \right\}.$$

Again we can change the homogeneous coordinates so that L_0 is the $\mathbb{CP}^{N-1}$ at infinity. We may now find the exhaustive sequence H_k as in Step 1.

This achieves the proof in the case when $\overline{\Omega}^\infty \cap \Sigma_0 = \emptyset$.

The general case is now an easy consequence. Indeed, since $\mathbb{CP}^n \setminus \Sigma_0$ is Stein, there is a strictly plurisubharmonic exhaustion function ψ. The sets

$$\Omega_c = \{\psi < c\}$$

form an exhaustive strongly pseudoconvex family for $\mathbb{CP}^n \setminus \Sigma_0$. Thus, in view of $(\star)$, there exists $\bar{c}$ such that $\overline{M} \subset \Omega_{\bar{c}}$. Up to a regularization of the boundary, $\Omega' \doteq \Omega \cap \Omega_{\bar{c}}$ is a strongly pseudoconvex open set satisfying $(\star)$, whose boundary contains M. Thus M can be extended thanks to what was already proved. $\quad\square$

Chapter 3
Semi-local extension by complex varieties

3.1. Introduction

The results in Chapter 2 deal with the global situation of submanifolds contained in the boundary of a special class of pseudoconvex unbounded domains in $\mathbb{C}^n$. In this chapter we deal with the boundary problem for complex analytic varieties in a "semi-local" setting. Throughout the chapter, we are going to apply some of the theorems of Chapter 2; we shall indifferently refer to it or to [16], which is its source.

More precisely, let $\Omega \subset \mathbb{C}^n$ be a strongly pseudoconvex open domain in $\mathbb{C}^n$, and $b\Omega$ its boundary. Let M be a maximally complex $(2m + 1)$-dimensional real closed submanifold ($m \geq 1$) of some open domain $A \subset b\Omega$, and let K be the boundary of A. We want to find a domain $\tilde{A}$ in Ω, independent from M, and a complex subvariety W of $\tilde{A}$ such that:

(i) $b\tilde{A} \cap b\Omega = A$;
(ii) $bW \cap b\Omega = M$,

In this chapter we show that, if $A \Subset b\Omega$, the problem we are dealing with has a solution $(\tilde{A}, W)$ whose $\tilde{A}$ can be determined in terms of the envelope $\widehat{K}$ with respect to the algebra of functions holomorphic in a neighborhood of $\overline{\Omega}$, i.e. we have the following

Th.: *For any maximally complex $(2m + 1)$-dimensional closed real submanifold M of A, $m \geq 1$, there exists an $(m + 1)$-dimensional complex variety W in $\Omega \setminus \widehat{K}$, with isolated singularities, such that $bW \cap (A \setminus \widehat{K}) = M \cap (A \setminus \widehat{K})$.*

(See Theorem 3.2.1).

If A is not relatively compact, this result can be restated in terms of "principal divisors hull", leading to a global result for unbounded strictly pseudoconvex domains, different from the results in Chapter 2. Indeed, this method of proof allows us to drop the Lupacciolu hypothesis ($\star$) and extend the maximally complex submanifold to a domain, which can anyhow be not the whole Ω. If the Lupacciolu hypothesis holds, then the domain of extension is in fact all of Ω. So this result is actually a generalization of the one in [16].

The crucial question of the maximality of the domain $\widetilde{A}$ we construct is not answered; in some simple cases the domain is indeed maximal.

In the last section, by the same methods, the extension result is proved for analytic sets (see Theorem 3.4.1).

It worths noticing that in [55] related results are obtained via a bump Lemma and cohomological methods. That approach may be generalized to complex spaces.

3.2. Main result

Let $\Omega \subset \mathbb{C}^n$ be a strongly pseudoconvex open domain in $\mathbb{C}^n$. Let A be a relatively compact subdomain of $b\Omega$, and $K = bA$. For any Stein neighborhood Ω_α of Ω we denote by $\widehat{K}_\alpha$ the hull of K with respect to the algebra of holomorphic functions of Ω_α, i.e.

$$\widehat{K}_\alpha = \{x \in \Omega_\alpha \; : \; |f(x)| \leq \|f\|_K \; \forall f \in \mathcal{O}(\Omega_\alpha)\} \,.$$

We define the hull $\widehat{K}$ as the intersection of the $\widehat{K}_\alpha$ when Ω_α varies through the family of all Stein neighborhoods of Ω. Observe that, since Ω is strongly pseudoconvex (and thus admits a fundamental system of Stein neighborhoods [63]), $\widehat{K}$ coincides with the hull of K with respect to the algebra of the functions which are holomorphic in some neighborhood of $\overline{\Omega}$, i.e.

$$\widehat{K} = \left\{x \in \overline{\Omega} \; : \; |f(x)| \leq \|f\|_K \; \forall f \in \mathcal{O}(\overline{\Omega})\right\}$$

(see [58]).

We claim that the following result holds:

Theorem 3.2.1. *For any maximally complex $(2m+1)$-dimensional closed real submanifold M of A, $m \geq 1$, there exists an $(m + 1)$-dimensional complex variety W in $\Omega \setminus \widehat{K}$, with isolated singularities, such that $bW \cap (A \setminus \widehat{K}) = M \cap (A \setminus \widehat{K})$.*

Following the same strategy as in [16] we first have a local extension result (see Lemma 3.2.2 below). In order to "globalize" the extension the main differences with respect to [16] are due the fact that we have to cut Ω with level-sets of holomorphic functions instead of hyperplanes. This creates some additional difficulties: first of all it is no longer possible to use the parameter which defines the level-sets as a coordinate. This can be overcome by working in $\mathbb{C}^{n+1}$ rather than in $\mathbb{C}^n$. Secondly, the intersections of tubular domains (see Lemmas 3.2.8, 3.2.11 and 3.2.12) may not be connected; since the extension is first achieved on each one of these tubular domains, this complicates the proof of the fact that different extension agree on the intersection.

With the same proof as in [16] we have:

Lemma 3.2.2. *There exist a tubular neighborhood I of A in Ω and an $(m + 1)$-dimensional complex submanifold with boundary $W_I \subset \overline{\Omega} \cap I$ such that $S \cap bW_I = M$.*

The hypothesis on $\widehat{K}$ allows us to prove the following:

Lemma 3.2.3. *Let $z^0 \in \Omega \setminus \widehat{K}$. Then there exist an open Stein neighborhood $\Omega_\alpha \supset \Omega$ and $f \in \mathcal{O}(\Omega_\alpha)$ such that*

1) $f(z^0) = 0$;
2) $\{f = 0\}$ *is a regular complex hypersurface of $\Omega_\alpha \setminus \widehat{K}$;*
3) $\{f = 0\}$ *intersects M transversally in a compact manifold.*

Moreover, if f is such a function for z^0, for any point z' sufficiently near to z^0, $f(z) - f(z')$ satisfies conditions 1), 2) *and* 3) *for z'.*

Proof. By definition of $\widehat{K}$, since $z^0 \in \Omega \setminus \widehat{K}$ there is a Stein neighborhood Ω_α such that $z^0 \notin \widehat{K}_\alpha$. So we can find a holomorphic function g in Ω_α such that $g(z^0) = 1$ and $\|g\|_K < 1$; $h(z) = g(z) - 1$ is a holomorphic function whose zero set does not intersect $\widehat{K}$. Since regular level sets are dense, by choosing a suitable small vector v and redefining h as $h(z + v) - h(z^0 + v)$ we can safely assume that h satisfies both 1) and 2).

We remark that $\{h = 0\} \cap b\Omega \Subset A$ by Alexander's theorem (see [1, Theorem 3]), and this shows compactness. Then, we may suppose that M is not contained in $\{z_1 = z_1^0\}$ and, for ε small enough, we consider the function $f(z) = h(z) + \varepsilon(z_1 - z_1^0)$. It's not difficult to see (by applying Sard's Lemma) that 3) holds for a generic ε. □

Now, we divide the proof of Theorem 3.2.1 in two cases, $m \geq 2$ and $m = 1$, observing that in the latter case proving that we can apply Harvey-Lawson to $\{f = 0\} \cap M$ is not automatic.

3.2.1. Dimension greater than or equal to 5: $m \geq 2$

For any $z^0 \in \Omega \setminus \widehat{K}$, Lemma 3.2.3 provides a holomorphic function such that the level $f_0 = \{f = 0\}$ contains z^0 and intersects M transversally in a compact manifold M_0. The intersection is again maximally complex (it is the intersection of a complex manifold and a maximally complex manifold, see [29]), so we can apply Harvey-Lawson theorem to obtain a holomorphic chain W_0 such that $bW_0 = M_0$. For τ in a small neighborhood $U \ni 0 \in \mathbb{C}$, the hypersurface $\{f_\tau = f - \tau = 0\}$ intersects M transversally along a compact submanifold M_τ which, again by Harvey-Lawson theorem, bounds a holomorphic chain W_τ. Observe that since $M_\tau \subset \{f_\tau = 0\}$, we have

$$W_\tau \subset \{f_\tau = 0\}.$$

We claim the following proposition holds:

Proposition 3.2.4. *The union $W_U = \bigcup_{\tau \in U} W_\tau$ is a complex variety contained in the open set $\widetilde{U} = \bigcup_{\tau \in U} f_\tau$.*

We need some intermediate results. Let us consider a generic projection $\pi : \widetilde{U} \to \mathbb{C}^m$ and set $\mathbb{C}^n = \mathbb{C}^{m+1} \times \mathbb{C}^{n-m-1}$, with holomorphic coordinates

$$(w', w), \, w' \in \mathbb{C}^{m+1}, \, w = (w_1, \ldots, w_{n-m-1}) \in \mathbb{C}^{n-m-1}.$$

Let $V_\tau = \mathbb{C}^{m+1} \setminus \pi(M_\tau)$, and $V = \bigcap_\tau V_\tau$.
For $\tau \in U$, $w' \in \mathbb{C}^{m+1} \setminus \pi(M_\tau)$ and $\alpha \in \mathbb{N}^{n-m-1}$, we define

$$I^\alpha(w', \tau) \doteq \int_{(\eta', \eta) \in M_\tau} \eta^\alpha \omega_{BM}(\eta' - w'),$$

where ω_{BM} is the Bochner-Martinelli kernel.

With the same proof of Lemma 2.4.3 we have the following:

Lemma 3.2.5. *Let $F(w', \tau)$ be the multiple-valued function which represents W_τ on $\mathbb{C}^{m+1} \setminus \pi(M_\tau)$ and denote by $P^\alpha(F(w', \tau))$ the sum of the αth powers of the values of $F(w', \tau)$. Then*

$$P^\alpha(F(w', \tau)) = I^\alpha(w', \tau).$$

In particular, the cardinality $P^0(F(w', \tau))$ of $F(w', \tau)$ is finite.

Remark 3.2.6. Lemma 3.2.5 implies, in particular, that the functions $P^\alpha(F(w', \tau))$ are continuous in τ. Indeed, they are represented as integrals of a fixed form over a submanifold A_τ which varies continuously with the parameter τ.

Lemma 3.2.7. *$P^\alpha(F(w', \tau))$ is holomorphic in the variable $\tau \in U \subset \mathbb{C}$, for each $\alpha \in \mathbb{N}^{n-m-1}$.*

Proof. Let us fix a point $(w', \underline{\tau})$ such that $w' \notin A_{\underline{\tau}}$ (this condition remains true for $\tau \in B_\epsilon(\underline{\tau})$). Consider as domain of $P^\alpha(F)$ the set $\{w'\} \times B_\epsilon(\underline{\tau})$. In view of Morera's theorem, we need to prove that for any simple curve $\gamma \subset B_\epsilon(\underline{\tau})$,

$$\int_\gamma P^\alpha(F(w', \tau))d\tau = 0.$$

Let $\Gamma \subset B_\epsilon(\underline{\tau})$ be an open set such that $b\Gamma = \gamma$. By $\gamma * M_\tau$ $(\Gamma * M_\tau)$ we mean the union of M_τ along γ (along Γ), *i.e.* $\bigcup_{\tau \in \gamma}\{\tau\} \times M_\tau$. Note that these sets are submanifolds of $\mathbb{C} \times \mathbb{C}^n$. The projection $\pi : \Gamma * M_\tau \to \mathbb{C}^n$ on the second factor is injective and $\pi(\Gamma * M_\tau)$ is an open subset of M bounded by $\pi(b\Gamma * M_\tau) = \pi(\gamma * M_\tau)$. By Lemma 3.2.5 and Stokes theorem

$$
\begin{aligned}
\int_\gamma P^\alpha(F(w', \tau))d\tau &= \int_\gamma I^\alpha(w', \tau)d\tau \\
&= \int_\gamma \left(\int_{(\eta', \eta) \in M_\tau} \eta^\alpha \omega_{BM}(\eta' - w') \right) d\tau \\
&= \iint_{\gamma * M_\tau} \eta^\alpha \omega_{BM}(\eta' - w') \wedge d\tau \\
&= \iint_{\Gamma * M_\tau} d\left(\eta^\alpha \omega_{BM}(\eta' - w') \wedge d\tau \right) \\
&= \iint_{\Gamma * M_\tau} d\eta^\alpha \wedge \omega_{BM}(\eta' - w') \wedge d\tau \\
&= \iint_{\pi(\Gamma * M_\tau)} d\eta^\alpha \wedge \omega_{BM}(\eta' - w') \wedge \pi_* d\tau \\
&= 0.
\end{aligned}
$$

The last equality follows from the fact that in $d\eta^\alpha$ appear only holomorphic differentials, η^α being holomorphic. But since all the holomorphic differentials supported by $\pi\Gamma * M_\tau \subset M$ already appear in $\omega_{BM}(\eta' - w') \wedge \pi_* d\tau$ (due to the fact that M is maximally complex and contains only $m + 1$ holomorphic differentials) the integral is zero. $\square$

Proof of Proposition 3.2.4. From [30] it follows that each W_τ has isolated singularities[1]. So, let us fix a regular point $(w_0', w_0) \in f_{\tau_0} \subset \tilde{U}$. In a neighborhood of this point $W = W_U$ is a manifold, since the construc-

---

[1] There could be singularities coming up from intersections of the solutions relative to different connected components of M_τ. These singularities are analytic sets and therefore should intersect the boundary. This cannot happen and so also these singularities are isolated.

tion depends continuously on the initial data. We want to show that W is indeed analytic in $\widetilde{U}$.

Let us fix $j \in \{1, \ldots, n - m - 1\}$ and consider multiindexes α of the form $(0, \ldots, 0, \alpha_j, 0, \ldots, 0)$. Let P_j^α be the corresponding $P^\alpha(F(w', \tau))$. Observe that for any j we can consider a finite number of P_j^α (it suffices to use $h = P_j^0(F(w', \tau))$ of them). By a linear combination of the P_j^α with rational coefficients, we obtain the elementary symmetric functions $S_j^0(w', k), \ldots, S_j^{h_j}(w', \tau)$ in such a way that for any point $(w', w) \in W$ there exists $\tau \in U$ such that $(w', w) \in W_\tau$. Thus:

$$Q_j(w', w, \tau) = S_j^h(w', \tau) + S_j^{h-1}(w', \tau)w_j + \cdots + S_j^0(w', \tau)w_j^h = 0.$$

In other words,

$$W \subset V = \bigcup_{\tau \in U} \bigcap_{j=1}^{n-m-1} \{Q_j(w', w, \tau) = 0\}.$$

Define $\widetilde{V} \subset \mathbb{C}^n(w', w) \times \mathbb{C}(\tau)$ as

$$\widetilde{V} = \bigcap_{j=1}^{n-m-1} \{Q_j(w', w, \tau) = 0\}$$

and

$$\widetilde{W} = W_\tau * U \subset \widetilde{V}.$$

Observe that, since the functions S_j^α are holomorphic, $\widetilde{V}$ is a complex subvariety of $\mathbb{C}^n \times U$. Since $\widetilde{V}$ and $\widetilde{W}$ have the same dimension, in a neighbourhood of (w_0', w_0, τ) $\widetilde{W}$ is an open subset of the regular part of $\widetilde{V}$, thus a complex submanifold. We denote by $Reg\,(\widetilde{W})$ the set of points $z \in \widetilde{W}$ such that $\widetilde{W} \cap \mathcal{U}$ is a complex submanifold in a neighbourhood $\mathcal{U}$ of z. It is easily seen that $Reg\,(\widetilde{W})$ is an open and closed subset of $Reg\,(\widetilde{V})$, hence a connected component. Observing that the closure of a connected component of the regular part of a complex variety is a complex variety we obtain the that $\widetilde{W}$ is a complex variety, $\widetilde{W}$ being the closure of $Reg\,(\widetilde{W})$ in $\widetilde{V}$.

Finally, since the projection $\pi : \widetilde{W} \to W$ is a homeomorphism and so is proper, it follows that W is a complex subvariety as well. $\square$

The varieties $\widetilde{W}_U$ that we have found patch together in such a way to define a complex variety on the whole $\Omega \setminus \widehat{K}$.

Lemma 3.2.8. Let $\widetilde{U}_f$ and $\widetilde{U}_g$ be two open subsets as in Proposition 3.2.4 and let W_f and W_g be the corresponding varieties. Let $z^1 \in \widetilde{U}_f \cap \widetilde{U}_g$. Then W_f and W_g coincide in a neighborhood of z^1.

Proof. Let $\lambda = f(z^1)$ and $\tau = g(z^1)$ and consider

$$L(\lambda', \tau') = \{f = h'\} \cap \{g = k'\} \subset \Omega$$

for (λ', τ') in a neighborhood of (λ, τ). Note that, for almost every (λ', τ'), $L(\lambda', \tau')$ is a complex submanifold of codimension 2 of $\widetilde{U}_f \cap \widetilde{U}_g$. Moreover, $W_f \cap L(\lambda', \tau')$ and $W_g \cap L(\lambda', \tau')$ are both solutions of the Harvey-Lawson problem for $M \cap L(\lambda', \tau')$ and consequently they must coincide. Since the complex subvarieties $L(\lambda', \tau')$ which are regular form a dense subset, W_f and W_g coincide on the connected component of $\widetilde{U}_f \cap \widetilde{U}_g$ containing z^1. $\qquad\square$

Remark 3.2.9. The above proof does not work in the case $m = 1$ since $M \cap L(\lambda', \tau')$ is generically empty.

In order to end the proof of Theorem 3.2.1, we have to show that the set S of the singular points of W is a discrete subset of $\Omega \setminus \widehat{K}$. Let $z^1 \in \Omega \setminus \widehat{K}$, and choose a function h, holomorphic in a neighborhood of Ω such that $h(z^1) = 1$ and $K \subset \{|h| \leq \frac{1}{2}\}$ and consider $f = h - \frac{3}{4}$. Observe that $z^1 \in \{\text{Re} f > 0\}$ and $K \subset \{\text{Re} f < 0\}$. Choose a defining function φ for $b\Omega$, strongly psh in a neighborhood of Ω and let us consider the family

$$(\phi_\lambda = \lambda(\varphi) + (1 - \lambda)\text{Re} f)_{\lambda \in [0,1]}$$

of strongly plurisubharmonic functions. For λ near to 1, $\{\phi_\lambda = 0\}$ does not intersect the singular locus. Let $\bar{\lambda}$ be the biggest value of λ for which $\{\phi_\lambda = 0\} \cap S \neq \emptyset$. Then the analytic set S touches the boundary of the Stein domain

$$\{\phi_{\bar{\lambda}} < 0\} \cap \Omega \subset \Omega.$$

So $\{\phi_{\bar{\lambda}} = 0\} \cap S$ is a set of isolated points in S. By arguing in the same way, we conclude that S consists of isolated points.

3.2.2. Dimension 3: $m = 1$

The first goal is to show that when we slice transversally M with complex hypersurfaces, we obtain maximally complex 1-dimensional real submanifolds.

Again, we fix our attention to a neighborhood of the form $\widetilde{U} = \cup_{\tau \in U} g_\tau$. Let us choose an arbitrary holomorphic $(1, 0)$-form ω in $\mathbb{C}^n$.

Lemma 3.2.10. *The function*

$$\Phi_\omega(\tau) = \int_{M_\tau} \omega$$

is holomorphic in U.

Proof. In view of Morera's theorem, we need to prove that for any simple curve $\gamma \subset U, \gamma = b\Gamma$,

$$\int_\gamma \Phi_\omega(\tau)d\tau = 0.$$

By Stokes theorem, we have

$$\int_\gamma \Phi_\omega(\tau)d\tau = \int_\gamma \left(\int_{M_k} \omega\right) d\tau$$

$$= \iint_{\gamma * M_\tau} \omega \wedge d\tau$$

$$= \iint_{\Gamma * M_\tau} d(\omega \wedge d\tau)$$

$$= \iint_{\Gamma * M_\tau} \partial\omega \wedge d\tau$$

$$= \iint_{\pi(\Gamma * M_\tau)} \partial\omega \wedge \pi_*d\tau$$

$$= 0.$$

The last equality is due to the fact that $\pi(\Gamma * M_\tau) \subset M$ is maximally complex and thus supports only $(2, 1)$ and $(1, 2)$-forms, while $\partial\omega \wedge \pi_*d\tau$ is a $(3, 0)$-form. $\qquad\square$

Lemma 3.2.11. *Let g be a holomorphic function on a neighborhood of Ω, and suppose*

$$\{|g| > 1\} \cap \widehat{K} = \emptyset.$$

Then there exists a variety W_g on $\Omega \cap \{|g| > 1\}$ such that

$$bW_g \cap b\Omega = M \cap \{|g| > 1\}.$$

Lemma 3.2.12. *Given two functions g_1 and g_2 as above, then W_{g_1} and W_{g_2} agree on $\{|g_1| > 1\} \cap \{|g_2| > 1\}$.*

In order to prove the previous Lemmas, we are going to use several times open subsets of the type $\widetilde{U}$ as in Proposition 3.2.4, so we need to fix some notation. Given an open subset $U \subset \mathbb{C}$, define $\widetilde{U}$ by

$$\widetilde{U} = \bigcup_{\tau \in U}\{f = \tau\}.$$

From now on we use open subsets $\widetilde{U}$ where $U = B(\overline{\tau}, \delta)$ is an open disc centered at $\overline{\tau}$ of radius δ. We say that $\{f = \overline{\tau}\}$ is the *core* of $\widetilde{U}$ and δ is its *amplitude*.

Proof of Lemma 3.2.11. For a fixed $d > 1$ consider the compact set $H_d = \overline{\Omega} \cap \{|g| \geq d\}$; we show that W_g is well defined on H_d. Let us fix also a compact set $C \subset \Omega$ such that W_I (see Lemma 3.2.2) is a closed submanifold in $H_d \setminus C$.

Consider all the open subsets $V_\alpha = \widetilde{U}_\alpha \cap \Omega$ (with $U_\alpha = B(\overline{\tau}_a, \delta_\alpha)$), constructed using only the function $f = g - 1$ up to addition of the function $\varepsilon(z_j - z_j^0)$ (see Lemma 3.2.3). If we do not allow ε to be greater than a constant $\overline{\varepsilon} > 0$, then by a standard argument of semicontinuity and compactness we may suppose that the amplitude of each $\widetilde{U}$ is greater than a constant δ.

We claim that it is possible to find a countable covering $\{U_i\}$ of H_d by a countable sequence V_i of those V_α in such a way to have

1. $V_0 \subset H_d \setminus C$;
2. if

$$B_l = \bigcup_{i=1}^{l} V_i$$

then $V_{l+1} \cap B_l \cap \Omega \neq \emptyset$.

The only thing to prove is the existence of V_0, since the second statement follows by a standard compactness argument.

Set $L = \max_{H_d} \mathsf{Re}\ g$. Since $\mathsf{Re}\ g$ is a non constant pluriharmonic function, $\{\mathsf{Re}\ g = L\}$ is a compact subset of $b\Omega \cap H_d$. Then we can choose $\eta > 0$ such that $\{\mathsf{Re}\ g = L - \eta\} \cap \Omega$ is contained in $H_d \setminus C$, and this allows us to define V_0.

Let $\widetilde{U}_1$ and $\widetilde{U}_2$ be two such open sets and $z^0 \in \widetilde{U}_1 \cap \widetilde{U}_2$. We can suppose that the cores of $\widetilde{U}_1$ and $\widetilde{U}_2$ contain z^0. They are of the form

$$f + \varepsilon_1(z_j - z_j^0) = \tau(\varepsilon_1), \quad f + \varepsilon_2(z_j - z_j^0) = \tau(\varepsilon_2).$$

For $\varepsilon \in (\varepsilon_1, \varepsilon_2)$, we consider the open sets $\widetilde{U}_\varepsilon$ whose core, passing by z^0, is $f + \varepsilon(z_j - z_j^0) = \tau(\varepsilon)$. Let I be the set of those ε satisfying the property: there exists an analytic subvariety $W_\varepsilon \subset \widetilde{U}_\varepsilon$ such that $W_\varepsilon = W_1 \cap \widetilde{U}_\varepsilon$. We must show that $I = (\varepsilon_1, \varepsilon_2)$.

I is open. Indeed, if $\varepsilon \in I$, then for ε' in a neighborhood of ε the core of $\widetilde{U}_{\varepsilon'}$ is contained in $\widetilde{U}_\varepsilon$ and so its intersection with M is maximally complex. Because of Lemma 3.2.10 the condition holds also for all the level sets in $\widetilde{U}_{\varepsilon'}$ and then we can apply again the Harvey-Lawson theorem [29] and the arguments of Proposition 3.2.4 in order to obtain $W_{\varepsilon'}$. Moreover, there is a connected component of $U_\varepsilon \cap U_{\varepsilon'}$ which contains z^0 and touches the boundary of Ω, where the W_ε and $W_{\varepsilon'}$ both coincide with

W_I (see Lemma 3.2.2). By virtue of the analytic continuation principle, they must coincide in the whole connected component.

I is closed. Indeed, since each $\widetilde{U}$ has an amplitude of at least δ, we again have that, for $\overline{\varepsilon} \in \overline{I}$, the intersection of $\widetilde{U}_{\overline{\varepsilon}}$ and $\widetilde{U}_{\varepsilon}$ must include (for $\varepsilon \in I$, $|\varepsilon - \overline{\varepsilon}|$ sufficiently small) a connected component containing z^0 and touching the boundary. We then conclude as in the previous case. $\qquad\square$

Proof of Lemma 3.2.12. Let us consider the connected components of $W_{g_1} \cap \{|g_2| > 1\}$. For each connected component W_1 two cases are possible:

1. W_1 touches the boundary of Ω: $W_1 \cap b\Omega \neq \emptyset$;
2. the boundary of W_1 is inside Ω:

$$bW_1 \Subset \{|g_1| = 1\} \cup \{|g_2| = 1\} \subset \Omega$$

In the former, the result easily follows in view of the analytic continuation principle (remember that on a strip near the boundary W_{g_1} and W_{g_2} coincide).

The latter is actually impossible. Indeed, suppose by contradiction that the component W_1 satisfies (2). Restrict g_1 and g_2 to W_1 and choose $t > 1$ such that

$$W_t \doteq \{|g_i| > t, \ i = 1, 2\} \Subset W_1.$$

The boundary bW_t of W_t consists of points where either $|g_1| = t$ or $|g_2| = t$. Choosing a point z_0 of the boundary where $|g_1| = t$ and $|g_2| > t$, then $|g_2|$ is a plurisubharmonic function on the analytic set

$$A = \{g_1 = g_1(z_0)\} \cap \{|g_2| \geq t\}.$$

Since $W_t \Subset W_1$, the boundary of the connected component of A through z_0 is contained in $\{|g_2| = t\}$. This is a contradiction, because of the maximum principle for plurisubharmonic functions. $\qquad\square$

3.3. Some remarks

3.3.1. Maximality of the solution

As stated above, we have not a complete answer to the problem of the maximality of $\widetilde{A}$. Nevertheless, here is a simple example where the constructed domain is actually maximal.

Example 3.3.1. Let $\Omega \subset \mathbb{C}^n$ be a strongly convex domain with smooth boundary, $0 \in \Omega$, and let h be a pluriharmonic function defined in a neighborhood U of $\overline{\Omega}$ such that $h(0) = 0$ and $h(z) = h(z_1, \ldots, z_{n-1}, 0)$ (*i.e.* h does not depend on z_n). Set

$$H = \{z \in U : h(z) = 0\}$$

and let

$$A = b\Omega \cap \{z \in U : h(z) > 0\}.$$

Then

$$\widetilde{A} = \Omega \cap \{z \in U : h(z) > 0\}.$$

In order to show that $\widetilde{A}$ is maximal for our problem, it suffices to find, for any $z \in H \cap \Omega$, a complex manifold $W_z \subset \widetilde{A}$ such that $M_z = \overline{W}_z \cap A$ is smooth and W_z cannot be extended through any neighborhood of z. We may suppose $z = 0$.

So, let $f \in \mathcal{O}(\overline{\Omega})$ be such that $\mathsf{Re}\, f = h$, $f(0) = 0$. We define

$$W_0 = \{z \in \widetilde{A} : z_n = e^{\frac{1}{f(z)}}\};$$

W_0 extends as a closed submanifold of $U \setminus \{f = 0\}$. Moreover, observe that each point of $\{f = 0\}$ is a cluster point of W_0. Suppose by contradiction that W_0 extends through a neighborhood V of 0 by a complex manifold W_0'; then $\{f = 0\} \cap V \subset W_0'$, thus $\{f = 0\} \cap V = W_0' \cap V$, which is a contradiction.

3.3.2. The unbounded case

Let $\Omega \subset \mathbb{C}^n$ be a strictly pseudoconvex domain, and $A \subset b\Omega$ an unbounded open subset of $b\Omega$.

Consider the set

$$\mathcal{A} = \left\{ A' \Subset b\Omega \mid A' \subset A,\ A'\ domain \right\}.$$

For an arbitrary $A' \in \mathcal{A}$ ($bA' = K'$), let $D_{A'}$ be the compact connected component of $\Omega \setminus \widehat{K'}$. Set

$$D = \bigcup_{A' \in \mathcal{A}} D_{A'}.$$

From Theorem 3.2.1 it follows that for every maximally complex closed $(2m + 1)$-dimensional real submanifold M of A, there is an $(m + 1)$-dimensional complex closed subvariety W of D, with isolated singularities, such that $bW \cap A = M$. So the domain D is a possible solution of our extension problem.

When $A = b\Omega$, we may restate the previous result in a more elegant way. In the same situation as above, consider

$$\mathbb{C}^n \subset \mathbb{CP}^n, \ \mathbb{C}^n = \mathbb{CP}^n \setminus \mathbb{CP}_\infty^{n-1}$$

and define the *principal divisors hull* $\widehat{C}_D$ of $C = \overline{\Omega} \cap \mathbb{CP}_\infty^{n-1}$ by

$$\widehat{C}_D = \left\{ z \in \Omega \mid \forall f \in \mathcal{O}(\overline{\Omega}) \ \{\widetilde{f = f(z)}\} \cap C \neq \emptyset \right\},$$

where $\{\widetilde{f = f(z)}\}$ is the closure of the connected component (in $\overline{\Omega}$) of $\{f = f(z)\}$ passing through z. Then

$$D = \Omega \setminus \widehat{C}_D.$$

Indeed, if $z \in D$, then there are an open subset $A' \subset b\Omega$ and a function $f \in \mathcal{O}(\overline{\Omega})$ such that $\{\widetilde{f = f(z)}\} \cap b\Omega$ is a compact submanifold of A'. In particular $z \notin \widehat{C}_D$. Vice versa, if $z \notin \widehat{C}_D$ then there is a function $g \in \mathcal{O}(\Omega')$ ($\Omega' \supset \Omega$ domain) such that $N = \{\widetilde{g = g(z)}\} \cap C = \emptyset$, i.e. N is a compact submanifold of $b\Omega$. Then, choosing a relatively compact open subset $A' \subset b\Omega$ large enough to contain N it follows that $z \in D_{A'} \subset D$.

3.4. Generalization to analytic sets

Let Ω, A and K be as before. We want now to consider the extension problem for analytic sets.

Let us recall that if $\mathcal{F}$ is a coherent sheaf on a domain U in $\mathbb{C}^n$, $x \in U$ and

$$0 \to \mathcal{O}_x^{m_k} \to \cdots \to \mathcal{O}_x^{m_0} \to \mathcal{F}_x \to 0$$

is a resolution of $\mathcal{F}_x$, then the *depth* of $\mathcal{F}$ at the point x is the integer $p(\mathcal{F}_x) = n - k$.

We will say that $M \subset A$ is a *k-deep trace* of an analytic subset if there are

i) an open set $U \subset \mathbb{C}^n$ such that $U \cap b\Omega = A$;
ii) an $(m + 1)$-dimensional irreducible analytic subset W_M of U such that $W_M \cap b\Omega = M$, whose ideal $\mathcal{I}_{W_M}$ has depth at least k at each point of U.

In this case, we say that the real dimension of M is $2m + 1$.

Theorem 3.4.1. *For any $(2m + 1)$-dimensional 4-deep trace of analytic subset $M \subset A$, there exists an $(m + 1)$-dimensional complex variety W in $\Omega \setminus \widehat{K}$, such that $bW \cap (A \setminus \widehat{K}) = M \cap (A \setminus \widehat{K})$.*

Observe that in this situation we already have a strip U on which the set M extends. So we only need to generalize Lemma 3.2.3 and the results in Section 3.2.1.

Lemma 3.4.2. *Let $z^0 \in \Omega \setminus \widehat{K}$. Then there exist an open Stein neighborhood $\Omega_\alpha \supset \Omega$ and $f \in \mathcal{O}(\Omega_\alpha)$ such that*

1. $f(z^0) = 0$;
2. $\{f = 0\}$ *is a regular complex hypersurface of $\Omega_\alpha \setminus \widehat{K}$;*
3. $\{f = 0\}$ *intersects M in a compact set and W_M in an analytic subset (of depth at least 3).*

Proof. The proof of the first two conditions is exactly the same as before. So, we focus on the third one.

Again, Alexander's theorem (see [1, Theorem 3]) implies compactness of the intersection with M. Then, we may suppose that W_M is not contained in $\{z_1 = z_1^0\}$ and, for ε small enough, let $f : \Omega_\alpha \to \mathbb{C}$ be the function $f(z) = h(z) + \varepsilon(z_1 - z_1^0)$, where Ω_α and h are as defined in Lemma 3.2.3. Consider the stratification of W_M in complex manifolds. By Sard's Lemma, the set of ε for which the intersection of $\{f(z) = 0\}$ with a fixed stratum is transversal is open and dense. Hence the set of ε for which the intersection of $\{f(z) = 0\}$ with each stratum is transversal is also open and dense, in particular it is non-empty. The conclusion follows. □

The previous Lemma enables us to extend each analytic subset

$$W_0 = W_M \cap \{f = 0\}$$

to an analytic set defined in the whole

$$\Omega \cap \{f = 0\}.$$

Indeed, on a strictly pseudoconvex corona the depth of W_0 is at least 3 and thus W_0 extends in the hole (see *e.g.* [6,54]). Obviously the extension lies in $\{f = 0\}$.

Observe that, up to an arbitrarily small modification of $b\Omega$ we can suppose that $b\Omega$ intersects transversally each stratum of the stratification of W_M. In this situation M is a smooth submanifold with negligible singularities of Hausdorff codimension at least 2 (see [21]).

Again, we consider a generic projection $\pi : \widetilde{U} \to \mathbb{C}^m$ and we use holomorphic coordinates (w', w), $w = (w_1, \ldots, w_{n-m-1})$ on

$$\mathbb{C}^n = \mathbb{C}^{m+1} \times \mathbb{C}^{n-m-1}.$$

Keeping the notations used in Section 3.2.1, let $V_\tau = \mathbb{C}^{m+1} \setminus \pi(M_\tau)$, and $V = \bigcap_\tau V_\tau$.

For $\tau \in U$, $w' \in \mathbb{C}^{m+1} \setminus \pi(M_\tau)$ and $\alpha \in \mathbb{N}^{n-m-1}$, we define

$$I^\alpha(w', \tau) \doteq \int_{(\eta', \eta) \in \text{Reg}(M_\tau)} \eta^\alpha \omega_{BM}(\eta' - w'),$$

ω_{BM} being the Bochner-Martinelli kernel.

Observe that the previous integral is well-defined and converges. Indeed, $W_\tau = W_M \cap \{f = \tau\}$ is an analytic set and thus, by Lelong's theorem, its volume is bounded near the singular locus. Hence, by Fubini's theorem, also the regular part of $M_\tau = W_\tau \cap b\Omega$ has finite volume up to a small modification of $b\Omega$.

Lemma 3.4.3. *Let $F(w', \tau)$ be the multiple-valued function which represents $\widetilde{M}_\tau$ on $\mathbb{C}^m \setminus \pi(M_\tau)$; then, if we denote by $P^\alpha(F(w', \tau))$ the sum of the αth powers of the values of $F(w', \tau)$, the following holds:*

$$P^\alpha(F(w', \tau)) = I^\alpha(w', \tau).$$

In particular, $F(w', \tau)$ is finite.

Proof. Let V_0 be the unbounded component of V_τ (where, of course, $P^\alpha(F(w', \tau)) = 0$). Following [29], it is easy to show that on V_0 also

$$I^\alpha(F(w', \tau)) = 0.$$

Indeed, if w' is far enough from $\pi(\text{Reg}(M_\tau))$, then $\beta = \eta^\alpha \omega_{BM}(\eta' - w')$ is a regular $(m, m-1)$-form on some Stein neighborhood Ω of $\text{Reg}(M_\tau)$. So, since in Ω there exists γ such that $\bar\partial \gamma = \beta$, we may write in the sense of currents

$$[\text{Reg}(M_\tau)](\beta) = [\text{Reg}(M_\tau)]_{m,m-1}(\bar\partial \gamma) = \bar\partial[\text{Reg}(M_\tau)]_{m,m-1}(\gamma).$$

We claim that $\bar\partial[\text{Reg}(M_\tau)]_{m,m-1}(\gamma) = 0$ and, in order to prove this, we first show that $[\text{Reg}(M_\tau)]$ is a closed current. Indeed, observe that $d[\text{Reg}(M_\tau)]$ is a flat current, since it is the differential of an L^1_{loc} current (see [24]). Moreover

$$S = \text{supp}(d[\text{Reg}(M_\tau)]) \subset \text{sing}(M_\tau),$$

hence, denoting by $\dim_\mathcal{H}$ the Hausdorff dimension and by $\mathcal{H}_s$ the s-Hausdorff measure, we have

$$\dim_\mathcal{H}(S) \leq \dim_\mathcal{H}(\text{sing}(M_\tau)) \leq \dim_\mathcal{H}(\text{Reg}(M_\tau)) - 2$$

and consequently that

$$\mathcal{H}_{\dim(\mathrm{Reg}(M_\tau))-1}(S) = 0.$$

By Federer's support theorem (see [24]), this implies that

$$d[\mathrm{Reg}(M_\tau)] = 0.$$

Now, since $\mathrm{Reg}(M_\tau)$ is maximally complex,

$$[\mathrm{Reg}(M_\tau)] = [\mathrm{Reg}(M_\tau)]_{m,m-1} + [\mathrm{Reg}(M_\tau)]_{m-1,m}.$$

Since $\overline{\partial}[\mathrm{Reg}(M_\tau)]_{m,m-1}$ is the only component of bidegree $(m, m-2)$ of $d[\mathrm{Reg}(M_\tau)]$ and $d[\mathrm{Reg}(M_\tau)] = 0$ we have

$$\overline{\partial}[\mathrm{Reg}(M_\tau)]_{m,m-1} = 0.$$

Moreover, since $[\mathrm{Reg}M_\tau](\beta)$ is analytic in the variable w', $[\mathrm{Reg}M_\tau](\beta) = 0$ for all $w' \in V_0$.

The rest of the proof goes as in Lemma 2.4.3.   □

Lemma 3.4.4. $P^\alpha(F(w', \tau))$ *is holomorphic in the variable* $\tau \in U \subset \mathbb{C}$, *for each* $\alpha \in \mathbb{N}^{n-m-1}$.

Proof. The only difference with the proof for the case of manifold is the fact that I is an integration performed over the regular part of $\Gamma \star M_\tau$ and not over $\Gamma \star M_\tau$. Nevertheless it is easy to see that Stokes theorem is valid also in this situation, so the chain of integrals in Lemma 3.2.7 holds in this case, too.   □

The rest of the proof of Theorem 3.4.1 goes as well as in the proof of Theorem 3.2.1 (see Section 3.2.1).

Chapter 4
Levi flat hypersurfaces in $\mathbb{C}^n$

As seen in the previous chapters, Levi flat hypersurfaces represent the obstruction of various extension problems. Therefore, we are often led to consider questions that involve the existence or the structure of Levi flat manifolds. This chapter is devoted to the most basic problem of the first kind, *i.e.* the boundary problem.

4.1. Boundaries of Levi flat hypersurfaces

The boundary problem for Levi flat hypersurfaces in $\mathbb{C}^n$ is a natural one in complex geometry, and has been studied quite extensively from as early as 1983, when Bedford and Gaveau [9] proved the following result: if a 2-sphere $S \subset \mathbb{C}^2$ is a graph over the boundary of a strongly pseudoconvex domain $D \subset \mathbb{C} \times \mathbb{R}$, and if it has only two complex points which are elliptic, then it is the boundary of a Levi flat graph over D which is also its hull of holomorphy. Subsequently this result has been improved in many directions, by dropping the graph hypothesis [10], by weakening the regularity assumptions [57], or by considering the unbounded case [58] (see also [2, 3] for the study of the polynomial hull of certain classes of compact subsets of $\mathbb{C}^2$, [26] for the case of toric boundaries, and [23] for the examination of related problems in an almost complex setting). However, the problem has been mainly studied for $n = 2$, taking as a starting point the classical result of Bishop [11] about the existence of 1-parameter families of holomorphic discs in the neighborhood of complex elliptic points of a hypersurface in $\mathbb{C}^2$. No such theorem exists in $\mathbb{C}^n$ for $n > 2$ and, in general, the situation is more complicated already from a local point of view. Indeed, it is easy to show that, contrarily to what happens in $\mathbb{C}^2$, a generic 2-codimensional hypersurface $S \subset \mathbb{C}^n$, $n > 2$, is not the local boundary of a Levi flat hypersurface, and compatibility conditions are required.

In what follows we give a brief account of the results contained in [22] where the problem of compact boundaries S of Levi-flat hypersurfaces of

$\mathbb{C}^n$, $n > 2$, is treated and sufficient conditions are given for a solution to exist. However, the solution is not given in term of a regular, embedded Levi-flat manifold but rather in terms of a Levi-flat chain which may have a singular set (for example, self-intersections). In Section 4.3 we will show that, if S is a graph contained in the boundary of a pseudoconvex domain, a regularity result can be proved in a quite simple way.

We begin with a discussion of the necessary conditions for the local existence of a Levi-flat manifold with boundary S.

4.1.1. Local conditions

Let $S \subset \mathbb{C}^n$ be a smooth, real submanifold of codimension 2. Then, for the holomorphic tangent space $H_p(S)$ at $p \in S$ the following holds:

$$n - 2 \leq \dim_{\mathbb{C}} H_p(S) \leq n - 1$$

i.e. either $H_p(S)$ has the least possible dimension - this is the generic situation - or $T_p(S)$ is a complex space. In the first case, we say that $p \in S$ is a *CR point*; clearly if $p \in S$ is a *CR* point then (since $\dim_{\mathbb{C}} H_p(S) = n - 2$ is an open condition) the points of a neighborhood of p are *CR* too, *i.e.* S is actually a *CR* submanifold near p. In the latter case, we say that p is a *complex point*. We separate the discussion of the necessary local conditions according to this distinction.

1. Condition at *CR* points

We restrict to a neighborhood of p in which S is generic, and suppose that there locally exist a Levi-flat hypersurface M such that $bM = S$. Since S is generic, it is transversal to the complex hypersurfaces of the foliation of M. Let M_p be the leaf of this foliation which passes through p; then $S_p = M_p \cap S$ is a real $(2n - 3)$-submanifold of S such that $H_q(S_p) = H_q(S)$ for each $q \in S_p$. It follows that S is *non-minimal* near p. This is a non-generic condition for S, which is trivial in $\mathbb{C}^2$ since in that case S is totally real at *CR* points - but this is no longer the case in $\mathbb{C}^n$ for $n > 2$.

In some special case it can also be seen that non-minimality is also a sufficient condition for S to be the local boundary of a Levi-flat manifold.

If all the *CR* orbits of S are 1-codimensional, then they are maximally complex submanifolds and thus they can be written as graphs of *CR* functions over hypersurfaces of $\mathbb{C}^{n-1}$. Those hypersurfaces are minimal because they are not Levi-flat (otherwise the *CR* orbits of S would not be of codimension 1), hence the extension theorems of Trépreau and Tumanov [64, 65] (see also Chapter 1) apply and the collection of holomorphic graphs which is obtained forms a local Levi-flat manifold with boundary S.

Another simple case is the following: if all the *CR* orbits are of codimension 2, then they are complex submanifolds by the Newlander-Nirenberg theorem; then S is clearly a locally flat boundary.

2. Condition at complex points

If $p \in S$ is a complex point, then $T_p(S)$ is a complex hyperplane. If we express, locally, S as a graph over its tangent plane we obtain, for a suitable choice of coordinates $(z_1, \ldots, z_{n-1}, w) = (z, w)$ in $T_p(S) \times \mathbb{C} \simeq \mathbb{C}^n$,

$$S = \{w = Q(z) + O(|z|^3)\}$$

where

$$Q(z) = \sum_{1 \le i,j, \le n-1} (a_{ij} z_i z_j + b_{ij} z_i \overline{z}_j + c_{ij} \overline{z}_i \overline{z}_j).$$

We say that S is *flat* at p if there exist coordinates (z, w) such that

$$\sum_{1 \le i,j, \le n-1} b_{ij} z_i \overline{z}_j \in \lambda \mathbb{R} \; \forall z \in \mathbb{C}^{n-1}$$

for some $\lambda \in \mathbb{C}$. In such a case, the same happens for all the choices of coordinates, since b_{ij} behaves as a tensor under biholomorphic maps.

Lemma 4.1.1. *Let $S \subset \mathbb{C}^n$ be a locally flat boundary near p; then S is flat at p.*

Proof. Let M be a local Levi-flat manifold such that $bM = S$.

- First of all, by "pushing" S slightly into M (*i.e.* choosing a C^2 approximation of S by submanifolds $\tilde{S}$ with a complex point) we may suppose $S \subset M$.
- Then (by considering the Taylor expansion of some function whose graph is M) we can approximate M up to the third order in z near p with a real analytic hypersurface M'; this does not affect the flatness property.
- As a last step, we change coordinates in such a way that

$$M' \to \{\mathsf{Im}w = 0\};$$

this is possible since M' is real analytic. In this case,

$$\sum_{1 \le i,j, \le n-1} b_{ij} z_i \overline{z}_j = \frac{Q(z) + Q(iz)}{2} \in \mathbb{R}$$

since $Q(z) \in \mathbb{R}$ for all $z \in \mathbb{C}^{n-1}$. $\qquad\square$

It is easy to see, by performing a coordinate change $(z, w) \to (z, \lambda w)$ in order to have $\sum b_{ij} z_i z_j \in \mathbb{R}$, and then another of the form $(z, w) \to (z, w + \sum a'_{ij} z_i z_j)$ in such a way that $a_{ij} + a'_{ij} = \bar{c}_{ij}$, that we can achieve $Q(z) \in \mathbb{R}$ when S is flat at p. When coordinates are chosen in this way, we say that S is in *flat normal form* at p.

3. Ellipticity

In order to give sufficient conditions for the existence of a (global) Levi flat manifold M with $bM = S$, we introduce a condition on elliptic points that is analogous to the one studied by Bishop in [11] for $n = 2$.

Definition 4.1.2. We say that S is *elliptic* at a flat point $p \in S$ if, in some flat normal form,

$$Q(z) \in \mathbb{R}^+ \text{ or } Q(z) \in \mathbb{R}^- \ \forall z \in \mathbb{C}^{n-1}.$$

In this case, the same thing happens in any flat normal form. Moreover, $b_{ij}(z) = (Q(z) + Q(iz))/2 \neq 0$, hence the matrix b_{ij} is definite.

This notion generalizes the ellipticity of real 2-dimensional submanifolds S of $\mathbb{C}^2$, in which case flatness is a trivial condition since the "mixed"part of the second order expansion reduces to the term $bz\bar{z}$. It can be shown that S is elliptic at p if and only if $L \cap S$ is elliptic (in Bishop's sense) for any complex 2-plane L such that $L \not\subset T_p(S)$.

Moreover, in a neighborhood of elliptic flat points it is possible to give a local result which is in some sense analogous to the classical Bishop's one:

Proposition 4.1.3. *Assume that S is nowhere minimal at its CR points, and has an elliptic flat points p. Then there exists a neighborhood V of p such that $V \setminus \{p\}$ is foliated by compact real $(2n - 3)$-dimensional CR orbits and there exists a Lipschitz function v, smooth and without critical points away from p, having the CR orbits as the level surfaces.*

Proof. (Sketch) We express S in flat normal form around p, *i.e.* $S = \{w = \varphi(z) = Q(z) + O(|z|^3)\}$ and we pose $S_0 = \{w = Q(z)\}$. Moreover, consider the unit ellipsoid $G = \{Q(z) = 1\} \subset \mathbb{C}^{n-1}$ and let π be the projection

$$\pi : \mathbb{C}^n \setminus \{z = 0\} \to G, \ (z, w) \to \frac{z}{\sqrt{Q(z)}}.$$

By a careful choice of coordinates, it is possible to prove that

$$H_{(z, \varphi(z))}(S) = H_{(z, Q(z))}(S_0) + O(|z|), \ z \neq 0, z \to 0$$

and

$$|z|\mathcal{L}_{(z,\varphi(z))}(S) = |z|\mathcal{L}_{(z,Q(z))}(S_0) + O(|z|), \ z \neq 0, z \to 0$$

where $|z|\mathcal{L}$ is the "normalized"Levi form. If $q \in S \setminus \{p\}$, denote by $E(q)$ the tangent space of the *CR* orbit through p. Since $E(q)$ is spanned by $H_q(S)$ and (the range of) the Levi form at q, which does not vanish in a neighborhood of p by the ellipticity hypothesis, $dim_{\mathbb{R}} H(q) = 2n - 3$. Denoting by $E_0(q)$ the analogous for S_0, we have

$$E(z, \varphi(z)) = E_0(z, Q(z)) + O(|z|), \ z \neq 0, z \to 0 \qquad (4.1.1)$$

because of the previous relations. Now, the following can be proved for the quadric S_0:

S_0 is CR and nowhere minimal outside p; the CR orbits are the $(2n-3)$-dimensional ellipsoids given by $\{w = const\}$. Moreover, the Levi form at the CR points is positive definite.

So, a *CR* orbit of S_0 has the form $\{w = Q(z) = c \in \mathbb{R}^+\}$, hence it is clear that

$$d\pi : E_0(q) \to T_{\pi(q)}(G)$$

is a bijection for all $q \in S_0 \setminus \{0\}$, *i.e.* the restriction of π to the *CR* orbits of S_0 is a diffeomorphism. This also implies, in view of 4.1.1, that in a small enough neighborhood of p the restriction of π to the *CR* orbits of S is a local diffeomorphism. A compactness argument allows then to prove that, for q sufficiently close to p, the *CR* orbit S_q through q is actually mapped diffeomorphically by π into G, hence it is compact. In this way is obtained a foliation of a neighborhood of p by global compact *CR* orbits.

In order to construct v, we choose any smooth curve $\gamma : [0, \varepsilon) \to S$ such that $\gamma(0) = p$; then the *CR* orbits S_q that are sufficiently close to p intersect γ in precisely one point $\gamma(t(S_q))$. We define v as

$$v(q) = \begin{cases} 0, & q = p; \\ t(S_q), & q \neq p. \end{cases}$$

this defines a smooth function outside p. To prove that v is Lipschitz at p, one observes that by 4.1.1 the derivatives of $E_{(z,\phi(z))}$ are $O(1/|z|)$, while the diameter of the orbits S_q is $O(|z|)$, which implies that the first derivatives of v are bounded near p. $\qquad \square$

This local foliation with compact CR orbits implies that, near elliptic flat points, S is a locally flat boundary at least in the weaker sense of locally being the boundary of an immersed Levi flat hypersurface M. This can be achieved by considering each CR orbit S_q - which is a maximally complex $(2n - 3)$-submanifold - as a graph of a CR function over its projection $\widetilde{S}_q$ over $H_p(S) \simeq \mathbb{C}^{n-1}$, and defining M as the family of the graphs of the relative holomorphic extensions.

4.1.2. Statement of the result

With the notions introduced in the previous section, we are able to state the main result contained in [22].

Theorem 4.1.4. *Let $S \subset \mathbb{C}^n$, $n > 2$, be a compact connected smooth real 2-codimensional submanifold such that the following hold:*

(i) *S is nonminimal at every CR point;*
(ii) *every complex point of S is flat and elliptic and there exists at least one such point;*
(iii) *S does not contain complex submanifolds of dimension $n - 2$.*

Then there exists a Levi flat $(2n - 1)$-subvariety $\widetilde{M} \subset \mathbb{C} \times \mathbb{C}^n$, with boundary $\widetilde{S}$ (in the sense of currents), such that the natural projection $\pi : \mathbb{C} \times \mathbb{C}^n \to \mathbb{C}^n$ restricts to a bijection which is a CR diffeomorphism between $\widetilde{S}$ and S outside the complex points of S.

As a matter of fact, the hypothesis of 4.1.4 give geometric constraints to S. We denote by S_{ell} the set of the elliptic flat points of S:

Proposition 4.1.5. *Let S satisfy the hypothesis of 4.1.4. Then S is homeomorphic to the unit sphere $S^{2n-2} \subset \mathbb{C}_z^{n-1} \times \mathbb{R}_x$ in such a way that the complex points are mapped to the poles $\{x = \pm 1\}$ and the CR orbits in S correspond to the $(2n - 3)$-spheres given by $\{x = const.\}$. In particular $\#S_{ell} = 2$ and $S_0 = S \setminus S_{ell}$ carries a foliation $\mathcal{F}$ of class C^∞ with 1-codimensional compact leaves.*

Proof. By hypothesis (i) and (ii) in 4.1.4, we can apply Proposition 4.1.3, thus obtaining that

- S_{ell} consists of isolated (hence finitely many) points;
- the CR orbits are diffeomorphic to the sphere $S^{2n-3} \subset \mathbb{R}^{2n-2}$, therefore they are simply connected.

Removing small saturated neighborhoods of the points in S_{ell}, we obtain a compact submanifold S_0 with boundary which, by hypothesis (iii), is foliated by 1-codimensional *CR* orbits in such a way that some of the leaves S_q satisfy $H^1(S_q, \mathbb{R}) = 0$, $\pi_1(S_q) = 0$.

If the foliation is transversely oriented (see [15]), then Thurston's Stability theorem [61] (or Reeb's one) allows to conclude that all the leaves are homeomorphic to S^{2n-3}, *i.e.*

$$S_0 \cong S^{2n-3} \times (0, 1)$$

which in turn implies that S satisfies the thesis.

If the foliation of S_0 is not transversely oriented, then the thesis can be obtained considering its transversely oriented 2-sheeted covering $\widetilde{S}_0$, with twice as many boundary leaves as S_0 (because they are simply connected). Then

$$\widetilde{S}_0 \cong S^{2n-3} \times (0, 1)$$

and the projection on S_0 is a homeomorphism on the leaves; this would induce a 2-sheeted covering of $[0, 1]$ over some 1-dimensional manifold, a contradiction. $\qquad\square$

4.1.3. Construction of the Levi flat chain

By Proposition 4.1.5, S is foliated by compact *CR* orbits which are maximally complex submanifolds. By Harvey-Lawson's theorem, every one of these *CR* orbits is the boundary of a complex subvariety. The natural candidate for M is the union of these subvarieties. In order to make this construction work, we must assure that the subvarieties vary smoothly , *i.e.* that their union actually forms a manifold (possibly up to a singular locus). This is achieved by a generalization of the results of [19] to the C^∞ case.

Let X be a complex manifold, and let $\mathcal{H}^d$ be the d-dimensional Hausdorff measure relative to some hermitian metric on X.

Definition 4.1.6. A closed subset $Y \subset X$ is called a *d-subvariety with negligible singularities* of class C^k if there exists a closed subset $\sigma \subset Y$, $\mathcal{H}^d(\sigma) = 0$, such that $Y \setminus \sigma$ is a closed, oriented, d-dimensional C^k submanifold of $X \setminus \sigma$ of locally finite $\mathcal{H}^d$ measure.

The minimal such σ is called the *singular set* of Y and $Reg\, Y = Y \setminus \sigma$ its *regular part*. Thanks to the assumption of locally finite $\mathcal{H}^d$ measure Y defines an integration current, which we denote again by Y. The usual *CR* notions - being *CR*, maximal complexity, holomorphicity, Levi flatness - extend to subvarieties with negligible singularities by requiring them to hold on the regular part (except, possibly, an additional $\mathcal{H}^d$-negligible closed subset).

The aim is now to use the function ν of Proposition 4.1.3 to map S to a submanifold S' of a real hyperplane $E \subset \mathbb{C}^{n+1}$, and apply the solution of the Harvey-Lawson problem with C^{∞} parameters to S' (whose maximally complex CR orbits are all contained in disjoint complex hyperplanes).

So, let $E \subset \mathbb{C}^n$ ($n \geq 4$) be a real hyperplane of equation $\{y_1 = 0\}$, and let k be the projection $k : E \to \mathbb{R}_{x_1}$. Let $N \subset E$ be a compact, closed (in the sense of currents) CR subvariety with negligible singularities, with singular locus τ, of real dimension $2n - 4$ and CR dimension $n - 3$. We shall assume the following:

(H) there exists a closed subset $\tau' \supset \tau$ of N, with $\mathcal{H}^{2n-4}(\tau') = 0$, such that, for every $z \in N \setminus \tau'$, $N \setminus \tau'$ is a submanifold transversal to the maximal complex affine subspace of E through z;

(H') there exists a closed subset $L_0 \subset \mathbb{R}_{x_1}$, with $\mathcal{H}^1(L_0) = 0$, such that for every $x_0 \in k(N) \setminus L_0$ the fiber $k^{-1}(x_0) \cap N$ is connected.

For $x_1 \in k(N)$, we set $N_{x_1} = N \cap E_{x_1} = N \cap k^{-1}(x_1)$ and define τ'_{x_1} as

$$\tau'_{x_1} = N_{x_1} \cap (\tau \cup \{z : E_{x_1} \text{ is not transverse to } N \text{ at } z\});$$

moreover, we pose

$$L = L_0 \cup \{x_1 \in \mathbb{R} : \mathcal{H}^{2n-5}(\tau'_{x_1}) > 0\}.$$

It is easy to see that

$$(\text{H}) \Rightarrow \mathcal{H}^1(L) = 0.$$

With these positions, we can state the following

Theorem 4.1.7. *Let N satisfy the hypothesis* (H) *and* (H')*, and let L be as above. Then, in $E' = E \setminus k^{-1}(L)$ there exists a unique C^{∞} maximally complex $(2n - 3)$-subvariety M with negligible singularities in $E' \setminus N$, foliated by complex $(n-2)$-subvarieties, with the property that M extends trivially to E' by a $(2n - 3)$-current (still denoted by M) such that $dM = \pm N$ in E'. The leaves are the sections by the hyperplanes E_{x_1}, $x_1 \in k(N) \setminus L$, and are the solutions of the Harvey-Lawson problem applied to $N \cap E_{x_1}$.*

The proof of Theorem 4.1.7 is quite involved. We give a brief survey of the important steps, many of which follow the proof given in [19] for the C^{ω} case.

Since N has CR dimension $n - 3$, its decomposition involve only term with bedegrees $(3, 1)$, $(2, 2)$ and $(1, 3)$. If $j : E \to \mathbb{C}^n$ is the immersion of the hyperplane E, there exists a closed, locally rectifiable current P such that $j_* P = N$. Then P decompose as follows:

$$P = P^{2,1} + P^{1,2} + dx_1 \wedge (P^{2,0} + P^{1,1} + P^{0,2})$$

where in this case the types are relative to the $(n-1)$-dimensional complex subspace of E. The following result holds on the components of P:

Lemma 4.1.8. *In a neighborhood of any point z^0, P is equal to a finite sum of distributions defined by C^∞ function in x_1. The same is true for each one of the components of P; we shall say that these components are C^∞ in x_1.*

If d'_E, d''_E are the d', d'' operators relative to $(z_2, \dots, z_n)$, $dP = 0$ implies

$$d''_E P^{1,1} + d'_E P^{0,2} = \frac{\partial P^{1,2}}{\partial x_1}, \quad d''_E P^{0,2} = 0.$$

Posing $E' = E \setminus k^{-1}(L)$, we look for a subvariety M such that $dM = N$ in E'. If M exists, there is a current T in $E' \setminus N$ such that $j_* T = M$; since M is maximally complex, the decomposition of T is

$$T = T^{1,1} + dx_1 \wedge (T^{1,0} + T^{0,1}).$$

Moreover, $dT = P$ implies

$$d''_E T^{1,0} + d'_E T^{0,1} = \frac{\partial T^{1,1}}{\partial x_1} - P^{1,1}$$

and

$$d''_E T^{1,1} = P^{1,2}, d''_E T^{0,1} = -P^{0,2}.$$

Lemma 4.1.9. *The equation $d''_E T^{0,1} = -P^{0,2}$, with $d''_E P^{0,2} = 0$, has a solution $U^{0,1}$ with compact support, C^∞ in x_1. All the solutions with compact support are qual to $U^{0,1}$ up to addition of $d''_E T$, where T is a current with compact support.*

The solution $U^{0,1}$ is given explicitly by means of convolution with suitably chosen kernels, and can be in turn used to express the coefficients of the rational function R introduced in the Harvey-Lawson article [29]. Let $\zeta' = (x_1, z_2, \dots, z_{n-1})$; then

$$C_m(\zeta') = K_E \sharp \pi_* [z_n^m (P^{1,1} + d'_E U^{0,1}) \wedge dx_1] \qquad (4.1.2)$$

where $\sharp$ is the convolution-contraction operator introduced in [29] and

$$K_E = \delta_0(z_2, \dots, z_{n-1}) \otimes H(x_1) \frac{\partial}{\partial x_1}$$

$H_1(x_1)$ being the Heaviside function in x_1. In fact, denoting by $<T,k,x'>$ the slice of a current T of E by $k^{-1}(x') = E_{x'}$, $x' \in \mathbb{R}_{x_1}$ (in our case the slice is just the restriction of the function), it can be shown that C_m defined as such satisfies

$$d'' < C_m(\zeta'), k, x' >=< z_n^m \pi_* P^{1,2}, k, x' >$$

where d'' is the usual operator on $E_{x'} \cong \mathbb{C}^{n-1}$. This equation shows that $< C_m(\zeta'), k, x' >$ is the same coefficient defined by Harvey and Lawson in [29]. This, along with (4.1.2), implies immediately

Proposition 4.1.10. *On $\mathcal{E} \setminus \mathcal{N} = \pi(E) \setminus \pi(N)$, for every $m \in \mathbb{N}$, the coefficient $C_m(\zeta')$ is a function C^∞ in x_1 and holomorphic in $z'' = (z_2, \ldots, z_{n-1})$.*

Then, again following [29], we define for $\zeta' \in \mathcal{E} \setminus \mathcal{N}$ and $w \in \mathbb{C}_{z_n} \setminus \overline{\Delta(0, \rho)}$ (for large enough ρ) the function

$$\varphi = C_0 \log w + \sum_{m=1}^{\infty} m^{-1} C_m w^{-m};$$

moreover we pose

$$R(\zeta', w) = \exp \varphi.$$

Denoting by $\{V_i\}_{i \in \mathbb{N}}$ the connected components of $\mathcal{E} \setminus \mathcal{N}$, we have

- R is C^∞ on $\mathcal{E} \setminus \mathcal{N} \times (\mathbb{C} \setminus \overline{\Delta(0, \rho)})$;
- $R \equiv 1$ on $V_0 \times (\mathbb{C} \setminus \overline{\Delta(0, \rho)})$;
- $R(\zeta', w) = \sum_{m=-\infty}^{C_0} A_m(\zeta') w^m$;
- every coefficient A_m is a polynomial in a finite number of C_m's.

Moreover, if $\mathcal{N}_0$ is a connected component of $\mathcal{N} \setminus Sing\mathcal{N}$, $\mathcal{N}_0 \subset \overline{V}_i \cap \overline{V}_j$, $\pi^{-1}(\mathcal{N}_0) \cap \mathcal{N}$ has $N_1, \ldots, N_r$ as connected components and

$$N_k = \{(\zeta', z_n) : \zeta' \in \mathcal{N}_0, z_n = \zeta_n^k(\zeta')\},$$

with ζ_n^l of class C^∞ in $\mathcal{N}_0$, then we have

$$R_i(\zeta', w) = \prod_{k=1}^{r} (w - \zeta_n^k(\zeta'))^{\pm} R_j(\zeta', w)$$

where R_i is the restriction of R to the closure of the domain V_i.

The previous expression allows to prove inductively that R is in fact a rational function (in the variable w) with coefficients holomorphic in $(z_2, \ldots, z_{n-1})$ and C^∞ in x_1.

The holomorphic chain in every E_{x_1} is then obtained, as in Harvey-Lawson's paper, by considering the divisor of the restriction of R to E_{x_1}. The C^∞ regularity of the coefficients of R in x_1 allows in the end to conclude the proof of 4.1.7.

Proof of Theorem 4.1.4. To conclude the proof observe, first of all, that by Propositions 4.1.3 and 4.1.5 we can construct a function $v : S \to \mathbb{R}$ such that

- v is C^∞ in $S \setminus S_{ell}$ and Lipschitz up to the elliptic points;
- the level sets of v are precisely the *CR* orbits of S.

Let

$$\widetilde{S} = \{(v(z), z) : z \in S\} \subset \mathbb{R} \times \mathbb{C}^n \subset \mathbb{C}^{n+1};$$

$\lambda(z) = (v(z), z)$ is a bicontinuous map $S \to \widetilde{S}$ and a *CR* diffeomorphism outside the elliptic points. Setting $N = \widetilde{S}$, N is a $(2n - 2)$-subvariety of $E = \varrho \times \mathbb{C}^n \subset \mathbb{C}^{n+1}$, with *CR* dimension $n - 2$ and with negligible singularities contained in $\lambda(S_{ell})$. Moreover, it is easy to see that N satisfies the hypothesis (H) and (H'). Hence, Theorem 4.1.7 applies, giving a Levi-flat $(2n - 2)$-variety $\widetilde{M} \subset \mathbb{R} \times \mathbb{C}^n$ such that $b\widetilde{M} = N$; taking the projection of $\widetilde{M}$ concludes the proof of 4.1.4. $\qquad\square$

4.2. Parameters

The aim of this section is to obtain a version with C^∞ parameter of the result contained in Chapter 2. We shall pose the question in the same way as Section 4.1.3, but we will also impose that the results in Chapter 2 apply, *i.e.* we will assume Lupacciolu's hypothesis $(\star)$.

Then, let us fix in $\mathbb{C}^n$ a coordinate system $(z_1, \ldots, z_n) = (z', z_n)$, $z_j = x_j + iy_j$. Let $E \subset \mathbb{C}^n$ be the real hyperplane spanned by the coordinates of z' and x_n. For any subset $A \subset \mathbb{R}$, we denote by E_A the set $E \cap \{x_n \in A\}$; if $A = \{a\}$ is just a real number, then $E_a = E_{\{a\}}$ is a complex (affine) hyperplane.

Consider a domain $\Omega \subset E$, with boundary of class at least C^2. We assume that the following conditions hold:

- Ω has no *complex tangencies*, *i.e.* $H_p(b\Omega) \subsetneq H_p(E)$ for every $p \in b\Omega$ (equivalently, the *CR* dimension of $b\Omega$ is $n - 2$ at every point $p \in b\Omega$).
- Ω is *strongly pseudoconvex*, *i.e.* the Levi form of $b\Omega$ is positive definite at any $p \in b\Omega$.

Remark 4.2.1. The previous pseudoconvexity assumption makes sense. In fact, because of the first point the complex tangent space $H_p(b\Omega)$ is properly contained in some E_a, $a \in \mathbb{R}$; it follows that the bracket that defines the Levi form takes still value in E_a, which implies that the Levi form takes value in the one-dimensional space $T_p(b\Omega) \cap E_a$. This means that in the case of Ω the Levi form can actually be seen as real valued, and this allows to consider its signature.

So, suppose that $\Omega \subset E$ is a strongly pseudoconvex subdomain, and let $M \subset b\Omega$ be a real, smooth CR submanifold of dimension $2n - 4$ and CR dimension $n - 3$. In analogy with Section 4.1.3, we define the following condition:

(H) there exist a closed subset $\tau \subset \mathbb{R}$, $\mathcal{H}^1(\tau) = 0$, such that for every $a \in \mathbb{R} \setminus \tau$ the complex hyperplane E_a intersects M transversally in a connected submanifold

Moreover, we introduce the Lupacciolu condition in the context we are dealing with:

$(\star)$ if $\overline{M}^\infty$ denotes the closure of $M \subset \mathbb{C}^n \subset \mathbb{CP}^n$ in $\mathbb{CP}^n$, then there exists an algebraic hypersurface V such that $V \cap \overline{M}^\infty = \emptyset$.

We have the following:

Theorem 4.2.2. *Let M and Ω be as above, and suppose that hypotheses (H) and $(\star)$ hold. Then there exists a $(2n - 3)$-real submanifold W of $\Omega \cap E_{\mathbb{R}\setminus\tau}$, with negligible singularities (see Section 4.1.3), such that $bW = M$ in $E_{\mathbb{R}\setminus\tau}$ and W is foliated in $(n - 2)$-complex subvarieties. Moreover, the leaves of the foliation are the intersections $W \cap E_a$ for $a \in \mathbb{R} \setminus \tau$.*

Proof. For any $a \in \mathbb{R} \setminus \tau$, let $M_a = M \cap E_a$. By hypothesis (H), M_a is a real, connected $(2n - 5)$-submanifold of $E_a \cong \mathbb{C}^{n-1}$; moreover, since M has CR dimension $n - 3$, M_a is maximally complex.

Clearly $\Omega_a = \Omega \cap E_a$ is a strongly pseudoconvex subdomain of E_a, and (by intersection) hypothesis $(\star)$ holds for the closure of M_a in $\mathbb{CP}^{n-1}$. Thus, we are in position to apply Theorem 2.5.1, obtaining a $(n - 2)$-complex submanifold $W_a \subset \Omega_a$, with isolated singularities, such that $bW_a = M_a$. We must check that $W = \cup_{a\in\mathbb{R}\setminus\tau} W_a$ is a submanifold with negligible singularities.

In order to establish this fact, we will prove that, if z belongs to the regular part of a W_a, then there is a neighborhood U of z such that

$$U \cap \left(\bigcup_{a\in\mathbb{R}\setminus\tau} W_a \right)$$

is a regular submanifold. In such a case, then, the union of the W_a is a regular submanifold outside a closed subset $Sing W$; moreover, $Sing W \cap E_a$ has zero $\mathcal{H}^{2n-4}$-measure. It follows that $Sing W$ has zero $\mathcal{H}^{2n-3}$-measure, hence W is a subvariety with negligible singularities.

So, fix $a \in \mathbb{R} \setminus \tau$ and $z \in Reg W_a$, and consider the same Veronese map $V : E_a \to \mathbb{C}^N$ already considered in the proof of Theorem 2.5.1. Then we can find a complex hyperplane H of $\mathbb{C}^N$ such that $V(z) \in H$ and H intersects $V(M_a)$ transversally in a compact submanifold.

Clearly, for a' close enough to a, applying the same Veronese map V to $M_{a'}$ we have that H intersects also $V(M_{a'})$ along a compact submanifold. It is enough to check that $\cup_{a'} V(W_{a'}) \cap H$ is a subvariety with negligible singularities.

Following the lines of Chapter 2, we choose a generic projection $\pi :$ $\mathbb{C}^N \to \mathbb{C}^{n-2}$ and we consider coordinates (w', w'') in $\mathbb{C}^N = \mathbb{C}^{n-2} \times \mathbb{C}^{N-n+2}$, $w' = (w_1, \ldots, w_{n-2})$, $w'' = (w_{n-1}, \ldots, w_N)$ in such a way that $H = \{w_N = 0\}$. We also choose $\varepsilon > 0$ and an open subset $U \subset \mathbb{C}$ in such a way that for any $a' \in (a - \varepsilon, a + \varepsilon)$ and any $k \in U$ the complex hyperplane $H_k = \{z_N = k\}$ intersects $V(M_{a'})$ transversally in a compact submanifold $M_{a',k}$.

Observe that the methods of Section 4.1.3 (applied to the hyperplane H_k) allow to show that, for fixed k, there exists a maximally complex subvariety of H_k (with negligible singularities) with boundary $H_k \cap M$. However, we can also finish the proof by following the lines of Chapters 2 and 3: for $k \in U$, $w' \in \mathbb{C}^{n-2} \setminus \pi(M_{a',k})$ and $\alpha \in \mathbb{N}^{N-n+2}$, we define

$$I^\alpha(w', k, a') \doteqdot \int_{(\eta', \eta) \in M_{a',k}} \eta^\alpha \omega_{BM}(\eta' - w'),$$

ω_{BM} being the Bochner-Martinelli kernel. The following Lemma holds with the same proof as 2.4.3:

Lemma 4.2.3. *Let $F(w', k, a')$ be the multiple-valued function which represents $V(W_{a',k})$ on $\mathbb{C}^{n-2} \setminus \pi(M_{a',k})$; then, if we denote by $P^\alpha(F(w', k, a'))$ the sum of the α-th powers of the values of $F(w', k, a')$, the following holds:*

$$P^\alpha(F(w', k, a')) = I^\alpha(w', k, a').$$

In particular, $F(w', k, a')$ is finite.

Since $I^\alpha(w', k, a')$ is clearly C^∞ in the variable a', being the integral of a fixed smooth form on a family of submanifolds which vary smoothly in the same parameter, it follows that also $P^\alpha(F(w', k, a'))$ is C^∞ in a'.

Now, there exist (we provide the details of the proof of this statement in Chapter 3) a finite set of polynomials in the variables (w''), whose coefficients are linear combinations with rational coefficients of some of the P^α's, such that $V(W_{a',k})$ is union of irreducible components of the analytic subset determined by these polynomials. It follows that, since $V(W_{a,k})$ is indeed a manifold in a neighborhood of $V(z)$, the union of the $V(W_{a',k})$ - which is determined locally by the zero locus of a polynomial with coefficients of class C^∞ in a' - must be a submanifold. □

4.3. Regularity of the solution

In $\mathbb{C}^n$, $n \geq 3$, we consider coordinates $(z_1, \ldots, z_n) \doteq (z', z_n)$, $(z_j = x_j + i y_j)$. Let Ω be a bounded, strongly convex open subset of $\mathbb{C}^{n-1}_{z'} \times \mathbb{R}_{x_n}$, with smooth boundary S. Let $f : S \to \mathbb{R}$ be a smooth function, and let $S' \subset \mathbb{C}^n$ be the graph of f. In accordance with Section 4.1, we suppose that

- S' is non-minimal in its CR points;
- S' has exactly two complex points p_1 and p_2, which are elliptic;
- S' does not contain complex subvarieties.

As previously shown, assuming these hypotheses there exist a function $\rho : S' \to \mathbb{C}^n$ with the following properties:

- ρ is smooth in $S' \setminus \{p_1, p_2\}$ and Lipschitz up to the elliptic points;
- the restriction of ρ to each CR-orbit of S' is constant (therefore, the image of each orbit is a maximally complex submanifold of $\mathbb{C}^n \times \mathbb{C}$);
- there exist a Levi-flat variety with negligible singularities $L \subset \mathbb{C}^n \times \mathbb{C}$ such that $d[L] = [\Gamma(\rho)]$ where $\Gamma(\rho)$ is the graph of ρ and the boundary is taken in the sense of currents.

Our purpose is to prove the following:

Theorem 4.3.1. *Let S and f be as above. Then, there exist a function $F : \Omega \to \mathbb{R}$, smooth on $\overline{\Omega} \setminus \{p_1, p_2\}$ and Lipschitz up to the elliptic points, whose graph M is a Levi flat hypersurface of $\mathbb{C}^n$ such that $bM = S'$. Moreover, the leaves of the foliation of M are regular complex manifolds with boundary.*

Proof. Denoting by π the projection $\pi : \mathbb{C}^n \times \mathbb{C} \to \mathbb{C}^n$, our candidate shall obviously be $M = \pi(L)$ where L is defined in the discussion above.
  We are going to prove the theorem through some steps.

- Preliminary considerations of regularity

Let N be a single CR-orbit of the foliation of $S' \setminus \{p_1, p_2\}$. Then, N is a maximally complex $2n - 3$ real submanifold of $\mathbb{C}^n$. Observe that, by the construction carried out in Section 4.1.3, the projection of the leaf of the foliation of L that corresponds to $\Gamma(\rho|_N)$ is $1 - 1$, and it gives exactly the Harvey-Lawson solution Σ for the boundary N. Moreover, N is contained in $b(\Omega \times \mathbb{R}_{y_n})$ and, since Ω is a strongly convex domain of $\mathbb{C}^{n-1} \times \mathbb{R}$, $\Omega \times \mathbb{R}$ is a strongly pseudoconvex domain. Then, by Harvey-Lawson's theorem [29] (along with the precisation contained in [30]) follows that Σ is a complex variety with isolated singularities, without self-intersections.

Now let N_1 and N_2 be two distinct *CR*-orbits, and let Σ_1, Σ_2 be the corresponding leaves. We claim that the intersection of Σ_1 and Σ_2 is empty. In fact, since they are both $(n - 1)$-analytic subsets of $\mathbb{C}^n$, their intersection I should be either empty or an analytic subset of positive dimension. Then I would meet the boundary of Σ_1, Σ_2 since both are Stein spaces and thus I would meet S'. But (by uniqueness and the fact that S' is locally a Levi-flat boundary, *i.e.* boundary of a Levi-flat *manifold*, see Section 4.1) in a neighborhood of S' we have that M is a regular, smoothly foliated submanifold, therefore two leaves cannot intersect there.

Remark 4.3.2. In the previous discussion, we have only employed the fact that $\Omega \times \mathbb{R}$ is a strongly pseudoconvex domain and S' is contained in its boundary, without regarding the graph nature of S'. Of course, if S' is not a graph it is always possible for the leaves to have, in fact, isolated singularities.

• The solution is contained in a graph

Let $(\zeta', \xi) \in \Omega$, and let H be a complex line contained in $\mathbb{C}^{n-1} \times \{\xi\}$. Define $\widetilde{H} = H \times \mathbb{R}_{x_n}$, and consider $\widetilde{\Omega} = \widetilde{H} \cap \Omega$. Then $\widetilde{\Omega}$ is a strongly convex domain of $\widetilde{H} \cong \mathbb{C} \times \mathbb{R}$, and the restriction of g to $b\widetilde{\Omega}$ is a smooth real function whose graph is $\widetilde{S}' = S' \cap (\widetilde{H} \times \mathbb{R}_{y_n}) \subset \widetilde{H} \times \mathbb{R}_{y_n} \cong \mathbb{C}^2$. Consider, as previously defined, $M = \pi(L)$ and define $\widetilde{M} = M \cap (\widetilde{H} \times \mathbb{R}_{y_n})$. For a generic choice of H, $\widetilde{H} \times \mathbb{R}_{y_n}$ intersects transversally the leaves of M and thus, in such a case, $\widetilde{M}$ is the union of a family of 1-dimensional analytic subsets. Clearly, the boundary of a connected component of any such analytic set is contained in $\widetilde{S}'$. It follows that $\widetilde{M}$ is contained in the polynomial hull of $\widetilde{S}'$, therefore, by Shcherbina's result (see [57]), is contained in a graph over $\widetilde{\Omega}$.

• Regularity of the solution

Let $p = (\zeta', \xi) \in \Omega$ and H be as before, and consider a small neighborhood U of p. For $q \in U$, let H_q be the translated of H which passes through q. With the notation corresponding to the one employed above, we can state the following

Lemma 4.3.3. *For a small enough neighborhood* $V \subset U$ *of* p, *let* $\mathcal{P}_q$ *be the polynomial hull of* $\widetilde{S}'_q$ *in* $\widetilde{H}_q \times \mathbb{R}_{y_n}$, *and let*

$$\mathcal{P} = \bigcup_{q \in V} \mathcal{P}_q.$$

Then $\mathcal{P}$ *is the graph of a continuous function over* V.

Proof. Let $\overline{q}$ be a point in V, and let $\{q_n\}_{n \in \mathbb{N}}$ be a sequence of points such that $q_n \to \overline{q}$. Then, obviously, the sets $\widetilde{S}'_{q_n}$ converge to the set $\widetilde{S}'_{\overline{q}}$ in the Hausdorff metric as $n \to \infty$. Moreover, it is also clear that $\widetilde{\Omega}_{q_n} \to \widetilde{\Omega}_{\overline{q}}$ for $n \to \infty$. Then, by Lemma 2.4 in [57] follows that $\mathcal{P}_{q_n} \to \mathcal{P}_{\overline{q}}$, $n \to \infty$. Since every $\mathcal{P}_q$ is a continuous graph, this allows to prove easily that $\mathcal{P}$ is a continuous graph as a whole. $\qquad \square$

Now we can deduce that the solution is a manifold. In fact, each leaf of M is an analytic subset of codimension 1 and it is a well-known fact that whenever such a set is contained in a continuous graph it is indeed a regular manifold (and the same holds for the corresponding leaf in L); by the previous Lemma, this is - at least locally - the case. The following proof of this fact was communicated to us by Jean-Marie Lion (here is slightly modified):

Lemma 4.3.4. *Let* H *be the germ of an analytic set of codimension* 1 *in* $\mathbb{C}^n$, *which is contained in the graph of a continuous function* $f : \mathbb{C}^{n-1} \times \mathbb{R} \to \mathbb{R}$. *Then* H *is in fact the germ of a complex manifold.*

Proof. We may suppose $0 \in H$. Let $h \in \mathcal{O}_n$ be a non identically vanishing germ of holomorphic function such that $H = \{h = 0\}$. Let D_ε be the disc $\{z' = 0\} \cap \{|z_n| < \varepsilon\}$. Then, for $\varepsilon << 1$, we have either $H \cap D_\varepsilon = 0$ or $H \cap D_\varepsilon = D_\varepsilon$. The latter is not possible since D_ε is not contained in any graph over $\mathbb{C}^{n-1} \times \mathbb{R}$. It follows that H is z_n-regular. Denote by π the projection $\pi : \mathbb{C}^n \to \mathbb{C}^{n-1}$. Then, the local parametrization theorem for analytic sets implies that there exists $d \in \mathbb{N}$ such that

- for some neighborhood U of 0 in $\mathbb{C}^{n-1}$, there exists an analytic set $\Delta \subset U$ such that $H_\Delta = H \cap (D_\varepsilon \times (U \setminus \Delta))$ is a manifold;
- $\pi : H_\Delta \to U \setminus \Delta$ is a d-sheeted covering.

We claim that the covering $\pi : H_\Delta \to U \setminus \Delta$ is trivial. Otherwise, there would exist a close loop $\gamma : S^1 \to U \setminus \Delta$ and $1 < d' \leq d$ such that, defining $s_k : S_1 \to S_1$ as $s_k(\theta) = k\theta$, we have that the lift by π of $\gamma_{d'} = \gamma \circ s_{d'}$ (denoted by $\widetilde{\gamma}_{d'}$) is a loop but the lift of γ_k, $k < d'$, is not. Define $\alpha, \beta : S^1 \to \mathbb{R}$ by $\alpha = \mathsf{Re}\, z_n(\widetilde{\gamma}_{d'})$, $\beta = \mathsf{Im}\, z_n(\widetilde{\gamma}_{d'})$;

since α is continuous there exists $\overline{\theta} \in S_1$ such that $\alpha(\overline{\theta}) = \alpha(\overline{\theta}')$ where $\overline{\theta}' = \overline{\theta} + 2\pi/d'$. But then $\gamma_{d'}(\overline{\theta}) = \gamma_{d'}(\overline{\theta}')$, thus $\beta(\overline{\theta}) = \beta(\overline{\theta}')$ since by construction $\beta(\theta) = f(\gamma_{d'}(\theta), \alpha(\theta))$ for $\theta \in S^1$. Hence $\tilde{\gamma}(\overline{\theta}) = \tilde{\gamma}(\overline{\theta}')$, a contradiction.

Since $\pi : H_\Delta \to U \setminus \Delta$ is a trivial covering, we may define d holomorphic functions $\tau_1, \ldots, \tau_d : U \setminus \Delta \to \mathbb{C}$ such that H_Δ is union of the graphs of the τ_j's. Note that, by the parametrization theorem, the functions τ_j are continuous up to Δ and therefore they extend as holomorphic functions $\tau_j \in \mathcal{O}(U)$. The thesis will follow from the fact that all the τ_j coincide. Suppose, by contradiction, $\tau_1 \neq \tau_2$; then for some disc $D \subset U$ centered at 0 we have $\tau_1|_D \neq \tau_2|_D$ and then, up to shrinking D, $(\tau_1 - \tau_2)|_D$ vanishes only in 0. But by the graph hypothesis $\{\mathrm{Re}(\tau_1 - \tau_2) = 0\} \subset \{\tau_1 - \tau_2 = 0\} = \{0\}$, and this is not possible since $\tau_1 - \tau_2$ is holomorphic and thus an open map (whose image must include a segment of the imaginary axis). $\qquad\Box$

Then, we observe that in the points p for which the leaf passing through p is regular L is a Levi-flat manifold, as proved in [22], Lemma 4.10. Therefore, since by the preliminary consideration above the projection $\pi|_L$ is injective (because of the fact that two leaves cannot intersect), it follows that M is also a manifold.

• Conclusion

We have proved that M is a smooth manifold, contained in a continuous graph over Ω. Since M is a compact manifold, it is easy to prove that its projection over $\mathbb{C}_{z'}^{n-1} \times \mathbb{R}_{x_n}$ must indeed be the whole Ω; therefore M is the graph of a smooth graph over Ω. $\qquad\Box$

• Remark on the elliptic points

In [37] it is proved that, if $p \in S \subset \mathbb{C}^2$ is an elliptic point, the Levi-flat manifold which is locally bounded by S is indeed C^∞ near p (Bishop had only proved that it is Lipschitz). The notion of ellipticity in $\mathbb{C}^n$ is given in such a way that $p \in S$ is an elliptic point for S if and only if it is elliptic for each $L \cap S$, where L is any complex 2-plane in $\mathbb{C}^n$. A *careful* book-keeping of the estimates used in [37] shows that they are indeed uniform with respect of L; so, in such a way it is possible to prove that the graph M is smooth also up to p_1 and p_2.

Chapter 5
Analytic multifunctions

In the last chapters of the thesis we continue the study of Levi flat manifolds, by considering some problems regarding - more generally - some classes of foliated, unbounded manifolds. Generally speaking, we will suppose that these manifolds are contained in a domain bounded "in some direction", and our aim will be to prove triviality results for (possibly some leaf of) their foliation. Precisely, in Chapter 6 we suppose that a real hypersurface S is a graph of a bounded function ρ, and we prove that ρ is constant with respect of some variables; in Chapter 7, instead, we deal with foliated manifolds contained in $\mathbb{C} \times D$. It turns out that, in the case of Levi flat manifolds, those results are a more or less direct application of the Liouville theorem for analytic multifunctions.

If X is an arbitrary set, we denote by $\mathcal{P}(X)$ the set of the subsets of X; a *multifunction* from Y to X is simply a map $Y \to \mathcal{P}(X)$.

Analytic multifunctions were introduced by Oka [49] as a generalization of holomorphic functions. Hartogs had shown that holomorphic functions were in fact characterized by the property of their graphs, to have pseudoconvex complementary in $\mathbb{C}^2$. Oka choose this property as a definition for analytic multifunctions, *i.e.* $f : \mathbb{C} \to \mathcal{P}(\mathbb{C})$ is analytic if, defining $\Gamma(f) = \cup_{z \in \mathbb{C}} (z, f(z)) \subset \mathbb{C}^2$, then $\mathbb{C}^2 \setminus \Gamma(f)$ is pseudoconvex.

Afterwards, Slodkowski [59] isolated an important property of analytic multifunctions: if p is a plurisubharmonic function,

$$p'(z) = \max_{f(z)} p$$

is subharmonic in $\mathbb{C}$. If we consider multifunctions f taking values in $\mathcal{P}(\mathbb{C})$, this is actually a characterization of analyticity; in such a way, it can be adopted as an alternative definition which is meaningful also in higher dimensions and gives a broader collection of analytic multifunctions. Moreover, this property is the key to prove many of the most interesting properties of analytic multifunctions, for example, the Liouville theorem mentioned above.

In this chapter, we give a brief introduction to the theory of analytic multifunctions, following a paper by T.Ransford [51]. In his work, a third point of view is assumed: basically, analytic multifunctions are those objects which can be "built up" by starting with the family of rational functions and performing two kinds of operations (in Ransford's terms, they are the *multigauge* generated by rational functions). One of the merits of this approach is that it provides a simple, abstract way to prove the basic properties of analytic multifunctions, in particular the "constructive" ones.

5.1. Definition

Gauges and multigauges

Let $\mathcal{G}$ be a family of upper semicontinuous functions $X \to [-\infty, +\infty]$ where X is a Hausdorff space. Let $u : X \to [-\infty, +\infty]$ be an upper semicontinuous function:

1. we say that $u \in \mathcal{G}^{\downarrow}$ if there exists a decreasing sequence $\{u_n\}$ in $\mathcal{G}$ which converges pointwise to u;
2. we say that $u \in \mathcal{G}^{\uparrow}$, or that u has *local $\mathcal{G}$-supports*, if for all $x_0 \in X$ with $u(x_0) < +\infty$ there exists $v \in \mathcal{G}$ such that $v(x_0) = u(x_0)$ and $v \le u$ on a neighborhood of x_0.

Definition 5.1.1. We say that $\mathcal{G}$ is a *gauge* if $\mathcal{G}^{\uparrow} = \mathcal{G}^{\downarrow} = \mathcal{G}$.

Example 5.1.2. Let X be a domain in $\mathbb{C}$. The family $\mathcal{G}$ of all the upper semicontinuous functions $u : X \to [-\infty, +\infty]$ which are subharmonic on $\{u < +\infty\}$ is a gauge. In fact, $\mathcal{G}^{\downarrow} \subset \mathcal{G}$ because it is a well known fact that (pluri)subharmonic functions are closed under monotone decreasing convergence. On the other hand, $\mathcal{G}^{\uparrow} \subset \mathcal{G}$ because functions u which have local $\mathcal{G}$ supports satisfy, by hypothesis 2., the mean value inequality

$$u(x_0) \le \frac{1}{\pi \varepsilon^2} \int_{B_\varepsilon(x_0)} u \ \forall x_0 \in X$$

for ε small enough; this implies that u is subharmonic.

By passing through charts, it is easy to see that the same statement holds when X is a Riemann surface.

We want to extend this definition to the case of set-valued functions. We will assume that all our multifunctions take values on $\mathbf{k}(Y)$, the set of the compact subsets of a Hausdorff space Y. The relation that we have in mind is the following: an upper semicontinuous function $f : X \to$

$[-\infty, +\infty] = Y$ produces a multifunction by defining $K_u : X \to \mathbf{k}(Y)$ in the following way:

$$K_u(x) = [-\infty, u(x)].$$

We say that a multifunction $K : X \to \mathbf{k}(Y)$ is *upper semicontinuous* if, for each open subset $V \subset Y$, the set $\{x \in X : K(x) \subset V\}$ is open in X, and that it is *lower semicontinuous* if, for each closed set $T \subset Y$, the set $\{x \in X : K(x) \subset T\}$ is closed in X. We say that K is *continuous* if it is both upper and lower semicontinuous. Then, clearly, by passing through the correspondence introduced above the previous definition gives the appropriate notion of (semi)continuity for functions.

Remark 5.1.3. We observe that, if Y is a metric space, then a continuous multifunction as defined above is continuous as a function $X \to \mathbf{k}(Y)$ when $\mathbf{k}(Y)$ is equipped with the Hausdorff metric.

Now, let X, Y be Hausdorff spaces, and let $\mathcal{M}$ be a family of upper semicontinuous multifunctions $X \to \mathbf{k}(Y)$. In analogy with the case of functions, let $K : X \to \mathbf{k}(Y)$ be a multifunction:

1. we say that $K \in \mathcal{M}^{\downarrow}$ if there exists a decreasing sequence $\{u_n\}$ in $\mathcal{M}$ such that $K_n \downarrow K$, *i.e.* $\cap_n K_n(x) = K$ for all $x \in X$;
2. we say that $K \in \mathcal{M}^{\uparrow}$, or that K has *local $\mathcal{M}$-supports*, if for all $x_0 \in X$ and each $y_0 \in bK(x_0)$ there exists $L \in \mathcal{M}$ such that $y_0 \in L(x_0)$ and $L(x) \subset K(x)$ on a neighborhood of x_0.

Definition 5.1.4. We say that $\mathcal{M}$ is a *multigauge* if $\mathcal{M}^{\uparrow} = \mathcal{M}^{\downarrow} = \mathcal{M}$.

Clearly, a family of functions $u : X \to [-\infty, +\infty]$ is a gauge if and only if the family of the respective K_u's form a multigauge. Moreover, observe that (by definition) the intersection of an arbitrary set of (multi)gauges is still a (multi)gauge.

Definition 5.1.5. Let X, Y be Hausdorff space, and let $\mathcal{L}$ be a family of upper semicontinuous multifunctions $X \to \mathbf{k}(Y)$. The *multigauge generated by $\mathcal{L}$* is the intersection $\mathcal{M}$ of the (non-empty, since it includes the multigauge formed by all the multifunctions) set of all the multigauges containing $\mathcal{L}$.

We want now to state two lemmas on multigauges which will allow to show that some properties of the one we will be interested in (*i.e.* that of analytic multifunctions) are inherited by those enjoyed by the family of functions that generate it (*i.e.* the rational functions). Let $f : X \to X'$ be a continuous map; then an upper semicontinuous multifunction $K : X \to$

$\mathbf{k}(Y)$ gives an upper semicontinuous multifunction $K' : X' \to \mathbf{k}(Y)$ by composition. If $\mathcal{K}$ is a collection of multifunctions on X, then we pose

$$\mathcal{K} \circ f = \{K \circ f : K \in \mathcal{K}\}.$$

Lemma 5.1.6 (Pull-back lemma). *Let X, X', Y be Hausdorff spaces. Let $\mathcal{L}, \mathcal{L}'$ be families of upper semicontinuous multifunctions $X, X' \to \mathbf{k}(Y)$, and let $\mathcal{M}, \mathcal{M}'$ be the multigauges generated by $\mathcal{L}, \mathcal{L}'$ respectively. Let $f : X' \to X$ be continuous. Then*

$$(\mathcal{L} \circ f) \subset \mathcal{L}' \Rightarrow (\mathcal{M} \circ f) \subset \mathcal{M}'.$$

Let X, Y, Y' be Hausdorff spaces, and let $F : X \times Y \to \mathbf{k}(Y')$ be an upper semicontinuous map. Given a multifunction $K : X \to \mathbf{k}(Y)$, we obtain a multifunction $X \to \mathbf{k}(Y')$ in the following way:

$$F(\cdot, K)(x) = F(x, K(x)) = \bigcup_{y \in K(x)} F(x, y);$$

it is easy to verify that $F(\cdot, K)$ takes values in $\mathbf{k}(Y')$ and is upper semicontinuous. Given a family $\mathcal{K}$ of multifunctions $X \to \mathbf{k}(Y)$, we note

$$F(\cdot, \mathcal{K}) = \{F(\cdot, K) : K \in \mathcal{K}\}.$$

Lemma 5.1.7 (Push-forward lemma). *Let X, Y, Y' be Hausdorff spaces, and let $\mathcal{L}, \mathcal{L}'$ be families of upper semicontinuous multifunctions $X \to \mathbf{k}(Y)$, $X \to \mathbf{k}(Y')$ respectively. Let $\mathcal{M}, \mathcal{M}'$ be the multigauges generated by $\mathcal{L}, \mathcal{L}'$. Suppose that the upper semicontinuous map $F : X \times Y \to \mathbf{k}(Y')$ satisfies the following condition:*

$$bF(x, C) \subset F(x, bC) \ \forall x \in X, C \in \mathbf{k}(Y).$$

Then

$$F(\cdot, \mathcal{L}) \subset \mathcal{L}' \Rightarrow F(\cdot, \mathcal{M}) \subset \mathcal{M}'.$$

An inductive application of the previous lemma yields to a more general result: if $F : X \times Y_1 \times \ldots \times Y_n \to \mathbf{k}(Y')$ is upper semicontinuous and satisfies

$$bF(x, C_1, \ldots, C_n) \subset F(x, bC_1, \ldots, bC_n) \ \forall x \in X, C_j \in \mathbf{k}(Y_j)$$

then

$$F(\cdot, \mathcal{L}_1, \ldots, \mathcal{L}_n) \subset \mathcal{L}' \Rightarrow F(\cdot, \mathcal{M}_1, \ldots, \mathcal{M}_n) \subset \mathcal{M}'.$$

A corollary of this statement will lead to a simple proof of the fact that Slodkowski's characterization holds for analytic multifunctions defined by means of multigauges. For an upper semicontinuous function v : $X \times Y_1 \times \ldots \times Y_n \to [-\infty, +\infty]$ we note

$$\max v(x, C_1, \ldots, C_n) = \max\{v(x, y_1, \ldots, y_n) : y_j \in C_j\}$$

where $C_j \in \mathbf{k}(Y_j)$; moreover, given multifunctions $K_j : X \to \mathbf{k}(Y_j)$, we note

$$\max v(\cdot, K_1, \ldots, K_n)(x) = \max v(x, K_1(x), \ldots, K_n(x)).$$

Corollary 5.1.8. *Let* $X, Y_j, \mathcal{L}_j, \mathcal{M}_j, v$ *be as above and let* $\mathcal{G}$ *be a gauge on* X. *Suppose that* v *satisfies (when the left-hand side* $< +\infty$*)*

$$\max v(x, C_1, \ldots, C_n) = \max v(x, bC_1, \ldots, bC_n) \; \forall x \in X, C_j \in \mathbf{k}(Y_j).$$

If $\max v(\cdot, K_1, ..., K_n) \in \mathcal{G}$ *whenever* $K_j \in \mathcal{L}_j$, *then* $\max v(\cdot, K_1, ..., K_n) \in \mathcal{G}$ *whenever* $K_j \in \mathcal{M}_j$.

Analytic multifunctions

Let $\mathbb{C}_\infty = \mathbb{C} \cup \{\infty\}$ be the Riemann sphere, and let $U \subset \mathbb{C}_\infty$ be an open subset.

Definition 5.1.9. Let $\mathcal{R}(U)$ be the family of all the multifunctions K : $U \to \mathbf{k}(\mathbb{C}_\infty)$ which are represented by a rational function, *i.e.*

$$K(z) = \{q(z)\}, \; z \in U$$

for some rational function $q : U \to \mathbb{C}_\infty$. Let $\mathcal{A}(U)$ be the multigauge generated by $\mathcal{R}(U)$. Elements of $\mathcal{A}(U)$ are called *analytic multifunctions* on U.

Clearly, constant multifunctions belong to $\mathcal{R}(U)^\uparrow$, hence they are analytic.

Despite its global definition, analyticity is a local property for multifunctions:

Proposition 5.1.10. *Let* $K : U \to \mathbf{k}(\mathbb{C}_\infty)$ *be a multifunction and let* $\mathcal{U}$ *be an open cover of* U. *Then*

$$K \in \mathcal{A}(U) \Leftrightarrow K|_{U'} \in \mathcal{A}(U') \; \forall U' \in \mathcal{U}.$$

In particular, for any open subset $U' \subset U$, $\mathcal{A}(U)|_{U'} = \mathcal{A}(U')$.

The proof of the previous proposition is achieved by applying Lemma 5.1.6 to the inclusions $U' \hookrightarrow U$, and by observing that being a local $\mathcal{A}(U)$-support is clearly a local property.

Moreover, the pull-back and push-forward lemmas allow to prove immediately that $\mathcal{A}(U)$ is preserved under rational changes of coordinates (although, of course, much more is true):

Proposition 5.1.11. *Let q be a rational function. Then*

1. $(K \circ q) \in \mathcal{A}(q^{-1}(U))$
2. $q(K) \in \mathcal{A}(U)$

Proof. 1. follows from 5.1.6 and 2. from 5.1.7.   □

5.2. Properties

Constructions

First of all, we give a list of examples and sufficient conditions for a multifunction K to belong to $\mathcal{A}(U)$.

Theorem 5.2.1. *Let $f : U \to \mathbb{C}_\infty$ be a meromorphic function. Then the multifunction*

$$K(z) = \{f(z)\}, \ z \in U$$

belongs to $\mathcal{A}(U)$.

Proof. It follows from the convergence of the Taylor polynomials of f to f.   □

We say that $K : U \to \mathbf{k}(\mathbb{C}_\infty)$ has *local meromorphic selection* if for any $z_0 \in U$ and $w_0 \in bK(z_0)$ there exist a neighborhood N of z_0 and a meromorphic function $f : N \to \mathbb{C}_\infty$ such that

$$f(z_0) = w_0 \text{ and } f(z) \in K(z) \ \forall z \in N.$$

Corollary 5.2.2. *If an upper semicontinuous multifunction K has local meromorphic selection then it is analytic.*

Proof. In fact, by 5.2.1 follows $K \in \mathcal{A}(U)^\uparrow$.   □

Remark 5.2.3. We observe that the previous criterium is not necessary for analyticity. Indeed, it can be proven that the kth-*root multifunction*

$$K(z) = \{w \in \mathbb{C} : w^k = z\}, \ z \in \mathbb{C}$$

belongs to $\mathcal{A}(\mathbb{C})$ (it also can be extended as a multifunction of $\mathcal{A}(\mathbb{C}_\infty)$), but does not admit a local meromorphic selection in a neighborhood of 0.

Now, we list some standard operations on multifunctions that respect analyticity:

Proposition 5.2.4. *Let $K_1, K_2 \in \mathcal{A}(U)$, and define the multifunction $K_1 \cup K_2$ as*

$$(K_1 \cup K_2)(z) = K_1(z) \cup K_2(z), \ z \in U.$$

Then $K_1 \cup K_2 \in \mathcal{A}(U)$.

If $C \in \mathbf{k}(\mathbb{C})$, we denote by $\widehat{C}$ the polynomial hull of C.

Proposition 5.2.5. *Let $K \in \mathcal{A}(U)$, and suppose that it is $\mathbf{k}(\mathbb{C})$-valued. Define the multifunction $\widehat{K}$ as*

$$\widehat{K}(z) = \widehat{K(z)}, \ z \in U$$

Then $\widehat{K} \in \mathcal{A}(U)$.

Theorem 5.2.6. *Let $K_1, \ldots, K_n \in \mathcal{A}(U)$, and let $f \in \mathcal{O}(V)$, where $V \subset \mathbb{C}^n$ is an open subset. Assume that*

$$K_1(z) \times \cdots \times K_n(z) \subset V$$

for all $z \in U$. Then $f(K_1, \ldots, K_n) \in \mathcal{A}(U)$.

The previous result derives from the multi-variable push-forward lemma, whose hypothesis are satisfies because f is holomorphic, hence open. The following consequence of 5.2.6 shows that some arithmetic operations are possible in $\mathcal{A}(U)$:

Corollary 5.2.7. *Let $K_1, K_2 \in \mathcal{A}(U)$ and suppose that $K_1(z), K_2(z) \subset \mathbb{C}$ for all $z \in U$. Define $K_1 + K_2$, $K_1 \cdot K_2$ in the following way:*

$$(K_1 + K_2)(z) = \{w_1 + w_2 : w_1 \in K_1(z), w_2 \in K_2(z)\},$$

$$(K_1 \cdot K_2)(z) = \{w_1 w_2 : w_1 \in K_1(z), w_2 \in K_2(z)\}.$$

Then $K_1 + K_2$ and $K_1 \cdot K_2$ are analytic.

In a similar way it can be also proved that, if $K \in \mathcal{A}(U)$, then the multifunction defined by the convex hull of $K(z)$, $z \in U$, is analytic.

Testing by psh functions

Theorem 5.2.8. *Let U be an open subset of $\mathbb{C}_\infty$, $K_1, \ldots, K_n \in \mathcal{A}(U)$ and let V an open subset of $\mathbb{C}^n$ such that*

$$K_1(z) \times \cdots \times K_n(z) \subset V \ \forall z \in U.$$

Let $v : V \to [-\infty, +\infty)$ be a plurisubharmonic function. Then

$$\max v(K_1, \ldots, K_n)$$

is a subharmonic function on U.

Proof. Clearly, if $C_1, \ldots, C_n \in \mathbf{k}(V)$, we have

$$\max v(C_1, \ldots, C_n) = \max v(bC_1, \ldots, bC_n)$$

because the restriction of v to each variable is subharmonic. Thus v satisfies the hypothesis of Corollary 5.1.8. Let $\mathcal{G}$ be the gauge of upper semicontinuous subharmonic functions. It is clear that $\max v(K_1, \ldots, K_n) \in \mathcal{G}$ for $K_1, \ldots, K_n \in \mathcal{R}(U)$. From 5.1.8 we deduce the thesis. $\square$

Now we are going to state some results that are proved with the aid of Theorem 5.2.8.

Theorem 5.2.9. *Let* $f : U \to \mathbb{C}_\infty$ *be a function, and let*

$$K(z) = \{f(z)\}, \ z \in U.$$

Then $K \in \mathcal{A}(U)$ *if and only if* f *is meromorphic on* U.

Proof. If f is meromorphic, then K belongs to $\mathcal{A}(U)$ by Theorem 5.2.1. Vice versa, suppose that K is analytic; since the property is local, we can suppose that U is a small disc around 0 and that f takes value in $\mathbb{C}$ (note that f is upper semicontinuous since K is). Applying 5.2.8 with $v(w) = \pm Re w$ we see that Ref is harmonic. Let h be the holomorphic function such that $Reh = Ref$ and $h(0) = f(0)$; then, (posing $K'(z) = \{h(z)\}$) by testing

$$v(w_1, w_2) = Re(w_1 - w_2)^2$$

on (K, K'), we have that $u = Re(f - h)^2$ is subharmonic; moreover, since $f - h \in i\mathbb{R}$, we have $u \le 0$ and $u(0) = 0$, which implies that $u \equiv 0$. This in turn implies that also $Im \ f = Im \ h$, *i.e.* f is holomorphic. $\square$

If K is a compact, connected subset of $\mathbb{C}$, its *logarithmic capacity* cap K is defined by

$$-\log \operatorname{cap} K = \lim_{z \to \infty} (g(z) - \log|z|)$$

where $g(z)$ is the Green function for the unbounded component of $\overline{\mathbb{C}} \setminus K$ having singularity at ∞.

The following theorem was first proved by Aupetit and Zraïbi [7]:

Theorem 5.2.10 (Picard theorem). *Let* $K \in \mathcal{A}(\mathbb{C})$, *and suppose that*

$$\widehat{K}(z) \subset \mathbb{C} \setminus E \ \forall z \in \mathbb{C}$$

where E *is a set of positive logarithmic capacity. Then* $\widehat{K}$ *is constant.*

Proof. If F is a compact subset of E with positive capacity and such that $\mathbb{C}_\infty \setminus F$ is connected, we test $\widehat{K}$ with $v = -g$, where g is the Green's function on $\mathbb{C}_\infty \setminus F$ with pole at ∞. Since $u = \max v(\widehat{K})$ is subharmonic and < 0 on $\mathbb{C}$, it follows by Liouville's theorem (for subharmonic functions) that $u \equiv -c$ for a positive c. Hence, taking a disc D on $\mathbb{C} \setminus F$ such that $g < c$, we have that K takes values on $\mathbb{C} \setminus D$. By performing a rational change of coordinates (see 5.1.11) we can suppose that $\widehat{K}$ is contained in the unit disc. The rest of the proof goes as in 7.1.2. $\square$

Also the maximum principle can be proved for analytic multifunctions:

Theorem 5.2.11 (Maximum principle). *Let U be an open subset of $\mathbb{C}_\infty$, and let $K \in \mathcal{A}(U)$ and $C \in \mathbf{k}(U)$. Then*

1. *if $C \neq \mathbb{C}_\infty$, then $bK(C) \subset K(bC)$;*
2. *if $C = \mathbb{C}_\infty$, then $bK(C) \subset \bigcap_{z \in \mathbb{C}_\infty} K(z)$.*

To prove 1., suppose by contradiction that $w_0 \in bK(C) \setminus K(bC)$, $w_0 \neq \infty$. Then there exists $w_1 \in \mathbb{C} \setminus K(C)$ such that

$$|w_0 - w_1| < \text{dist}(w_1, K(bC)).$$

We test the multifunction K by

$$v(w) = \frac{1}{|w - w_1|};$$

since $u = \max w(K(z))$ is subharmonic by 5.2.8, we have

$$\max_{bC} u \leq \frac{1}{\text{dist}(w_1, K(bC))} < \frac{1}{|w_0 - w_1|} \leq \max_C u,$$

which contradicts the subharmonicity of u. The second point follows from the first one: if $w_0 \in bK(\mathbb{C}_\infty) \setminus K(z_0)$, by semicontinuity follows that there exists $C_0 \in \mathbf{k}(\mathbb{C})$ such that $w_0 \in bK(C_0) \setminus K(bC_0)$.

Finally, we state a result which says that the various approaches for the definition of analytic multifunctions (whose graph is contained in $\mathbb{C}^2$) are equivalent:

Theorem 5.2.12. *Let U be an open subset of $\mathbb{C}$, K an upper semicontinuous multifunction on U with values on $\mathbf{k}(\mathbb{C})$, and define the graph Γ of K to be*

$$\Gamma = \{(z, w) \in U \times \mathbb{C} : w \in K(z)\}.$$

The following are equivalent:

1. $K \in \mathcal{A}(U)$;
2. *given U' open in U and v psh on a neighborhood of $\Gamma \cap (U' \times \mathbb{C})$, the function $u(z) = \max v(z, K(z))$ is subharmonic on U';*
3. *the function*

$$(z, w) \to -\log \text{dist}(w, K(z))$$

is psh on $(U \times \mathbb{C}) \setminus \Gamma$;
4. *the set $(U \times \mathbb{C}) \setminus \Gamma$ is pseudoconvex.*

We remark that 1. $\Rightarrow$ 2. is contained in 5.2.8, while 2. $\Rightarrow$ 3. and 3. $\Rightarrow$ 4. are proved in [59] (see also Lemma 7.1.1 for a proof of 4. $\Rightarrow$ 2.). The proof of 4. $\Rightarrow$ 1. is achieved in roughly the following way: chosen a psh exhaustion function ρ for $(U \times \mathbb{C}) \setminus \Gamma$, we define K_n by

$$K_n(z) = K \cup \{w : \rho(z, w) \geq n\}.$$

It can be shown, by taking a suitable quadratic approximation of ρ, that K_n has local $\mathcal{A}(U)$-supports. Afterwards, since $K_n \downarrow K$, we obtain $K \in \mathcal{A}(U)$.

Chapter 6
On bounded Levi flat graphs

6.1. Statement

Consider, in $\mathbb{C}^2$, coordinates (z, w) with $z = x + iy$ and $w = u + iv$. We denote by π the projection $\pi : \mathbb{C}^2 \to \mathbb{C}_z$ and by τ the projection $\tau : \mathbb{C}^2 \to \mathbb{C}_z \times \mathbb{R}_u$. Let $\rho : \mathbb{C} \times \mathbb{R} \to \mathbb{R}$ be a smooth function such that

$$S = \{v = \rho(z, u)\}$$

is a Levi flat graph. Then the leaves of the foliation associated to S are regular (locally closed, embedded) complex manifolds of dimension 1. We claim that the following holds true:

Theorem 6.1.1. *Suppose that $|\rho|$ is bounded by some constant M; then S is foliated by complex lines, i.e. ρ has the form*

$$\rho(z, u) = \varrho(u)$$

for some smooth function $\varrho : \mathbb{R} \to \mathbb{R}$.

Actually, we are going to prove a sharper result. If S is any foliated 3-submanifold, we say that a leaf Σ of S is *properly embedded* if, for (almost) every ball $B \subset \mathbb{C}^2$, the connected components of $\overline{B} \cap \Sigma$ are compact, embedded submanifolds of $\overline{B} \cap S$ with boundary. We say that the foliation of S is *proper* if the leaves are properly embedded.

Theorem 6.1.2. *Let $S = \{v = \rho(z, u)\}$ be a properly foliated hypersurface of $\mathbb{C}^2$ (not necessarily a Levi flat one), and suppose that $|\rho|$ is bounded. Then every complex leaf of S is actually a complex line.*

Theorem 6.1.1 is a consequence of Theorem 6.1.2, since the foliation of a Levi flat graph is proper (see, for example, the main Theorem of [57]). However, it can be more easily proven by means of analytic multifunctions.

Proof of Theorem 6.1.1. By hypothesis there exists a complex line $\{w = c\}$ such that S lies outside the cylinder

$$C = \{(z, w) : |w - c| < \varepsilon\}.$$

Then, we can perform a rational change of coordinates (acting only on the w-coordinate) such that the image S' of S is contained in $\mathbb{C}_z \times D_w$, where D_w is the unit disc. The complementary of

$$\overline{S}' = S' \cup (\mathbb{C}_z \times \{0\})$$

in $\mathbb{C}_z \times D_w$ is pseudoconvex. Indeed, a psh exhaustion function φ for $\mathbb{C}^2 \setminus S$ induces a psh exhaustion function φ' for on $\mathbb{C}_z \times (D_w \setminus \{0\}) \setminus S'$; then

$$\psi = \max\left\{\varphi' \left|\frac{1}{w}\right|\right\}$$

is a psh exhaustion function for $\overline{S}'$ on $\mathbb{C}_z \times D_w$. The rest of the proof can be carried out in the same way as the one of Theorem 7.0.7, with some additional care due to the fact that, if f is the multifunction representing S', $f(z)$ is no longer (generically) a C^1 curve but only a C^0 one (though $f(z) \setminus \{0\}$ *is a* C^1 curve). □

On the other hand, Theorem 6.1.2 cannot be treated by applying directly the methods of analytic multifunctions, since the complement of S need not be pseudoconvex in that case.

6.2. Preliminary results

From now on we assume that S is as in 6.1.2. First of all, we prove the following

Lemma 6.2.1. *Let* Σ *be any complex leaf of the foliation of* S. *Then the projection* $\pi|_\Sigma$ *is a local homeomorphism.*

Proof. Let $p \in \Sigma$, $p = (p_z, p_w)$; it suffices to show that the differential of $\pi|_\Sigma$ is surjective in p. In the opposite case, there would exist neighborhoods U of p in $\mathbb{C}^2$ and V of p_w in $\mathbb{C}_w$, and a holomorphic function $f : V \to \mathbb{C}$ such that, denoting by Σ' the connected component of $\Sigma \cap U$ which contains p, we would have

$$\Sigma' = \{z = f(w)\}$$

and $\frac{\partial f}{\partial w}(p_w) = 0$. In other words, $\partial/\partial w \in T_p^\mathbb{C}(\Sigma)$ and thus $\partial/\partial v \in T_p(\Sigma)$. This would imply

$$\frac{\partial}{\partial v} \in T_p(S),$$

which contradicts the fact that ρ is a smooth function on $\mathbb{C}_z \times \mathbb{R}_u$. □

Lemma 6.2.1 shows that a complex leaf Σ of the foliation is locally a graph over $\mathbb{C}_z$. Anyway, since we do not know whether $\pi : \Sigma \to \mathbb{C}_z$ is actually a covering, we cannot conclude immediately that $\pi|_\Sigma^{-1}$ is single-valued. However, if this is the case, it is easy to deduce that the conclusion of Theorem 6.1.1 holds true for Σ, provided that the projection $\pi|_\Sigma$ is onto:

Lemma 6.2.2. *Let Σ be a complex leaf of S, and suppose that*

- $\pi(\Sigma) = \mathbb{C}_z$;
- *for every $z_0 \in \mathbb{C}_z$, $\pi^{-1}(z_0) \cap \Sigma$ is a single point.*

Then there exists $c \in \mathbb{C}$ such that

$$\Sigma = \{w = c\}.$$

Proof. Indeed, in this case the leaf Σ is biholomorphic to $\mathbb{C}$ as $\pi|_\Sigma$ is one to one; then, denoting by v the projection on the v-coordinate, $v \circ (\pi|_\Sigma)^{-1}$ is a harmonic, bounded function on $\mathbb{C}_z$, which is constant by Liouville's Theorem. Therefore $v|_\Sigma$ is also constant and so is $u|_\Sigma$, which is conjugate to v in Σ. $\qquad\square$

Remark 6.2.3. One may ask whether the latter hypothesis in Lemma 6.2.2 can be replaced by

- $\pi|_\Sigma$ is a local homeomorphism.

This is not the case. Here we present a relatively simple example, however, much more is true (Fornaess and Stout showed that any complex manifold is the image of a polydisc by a finite-to-one local biholomorphism).

Example 6.2.4. Consider the subset

$$L = \{(x, y) \in \mathbb{R}^2 : |y| < 1\}.$$

It is simple to show that there exists a map $\phi : L \to \mathbb{C}$ such that

- ϕ is onto;
- the differential of ϕ is always invertible;
- ϕ extends as a continuous function $\overline{L} \to \mathbb{C}$;
- $\phi^{-1}(z)$ consists of finitely many point for every $z \in \mathbb{C}$.

To be convinced of this fact, one may proceed as follows. A map of this kind can be identified with the smooth motion of an open segment on $\mathbb{C}$ along a curve parametrized by $\mathbb{R}_x$. We can, for example, first cover a ball $B \subset \mathbb{C}$ with this motion, and then let the segment proceed along a suitably chosen spiral to fill in the whole $\mathbb{C}$.

Then, denoting by J the standard complex structure on $\mathbb{C}$, we can endow L with the complex structure $\phi^*(J)$, thus obtaining a simply connected open Riemann surface $L_\mathbb{C}$ for which $\phi : L_\mathbb{C} \to \mathbb{C}$ is tautologically holomorphic. By Riemann's uniformization Theorem, we find a biholomorphism $\psi : L_\mathbb{C} \to X$ where X is either $\mathbb{C}$ or the unit disc $D \subset \mathbb{C}$. We claim that first case is not possible. Indeed, consider the set

$$A_\varepsilon = L_\mathbb{C} \cap D((0, 1), \varepsilon)$$

where $D((0, 1), \varepsilon)$ is a disc with center $(0, 1) \in \mathbb{R}^2$ and radius $\varepsilon \ll 1$. If ε is small enough, A_ε is mapped by ϕ biholomorphically on a open set of $\mathbb{C}$. Moreover, consider $\psi(A_\varepsilon) \subset \mathbb{C} \subset \mathbb{CP}^1$; observe that the boundary of $\psi(A_\varepsilon)$ contains $\{\infty\}$ and (for small ε) $0 \notin \psi(A_\varepsilon)$. If $\{\mathcal{U}_n\}$ is a fundamental system of neighborhoods of

$$\overline{A}_\varepsilon \cap \{y = 1\}$$

in $\overline{A}_\varepsilon$, it is also clear that the sets $\psi(\mathcal{U}_n)$ approach $\infty \in \mathbb{CP}^1$ as $n \to +\infty$. Let $g \in \mathcal{O}(\mathbb{CP}^1 \setminus \{0\})$ be such that $g(\infty) = 0$ and $g \not\equiv 0$. Then by what was observed above it follows that $f \in \mathcal{O}(\phi(A_\varepsilon))$ defined as $f = g \circ \psi \circ \phi^{-1}$ is continuous up to $\phi(\{y = 1\})$, and vanishes on this set. This is a contradiction.

It follows that $X = D$. Let $i : D \to \mathbb{C}^2$ be defined as

$$i(z) = (\phi \circ \psi^{-1}(z), z);$$

then i is a holomorphic embedding of D in $\mathbb{C}^2$ and we set $\Sigma = i(D)$. Observe that $\pi : \Sigma \to \mathbb{C}$ is onto and a local homeomorphism; moreover, by construction $|v| < 1$ on Σ. Anyway, obviously Σ cannot be the leaf of any foliation of a Levi-flat graph in $\mathbb{C}^2$.

In order to prove Theorem 6.1.1 our aim is to apply Lemma 6.2.2 and so, from now on, we shall focus on a single complex leaf Σ of the foliation of S and we will prove that its projection over $\mathbb{C}_z$ is a biholomorphism. We set

$$\pi(\Sigma) = \Omega \subset \mathbb{C}_z;$$

then, since Σ is a complex curve (or also because of Lemma 6.2.1), Ω is an open subset of $\mathbb{C}_z$.

We suppose, by contradiction, that $\Omega \subsetneq \mathbb{C}_z$, and let $z_0 \in b\Omega$. The following result shows that in fact z_0 must belong to Ω at least in some special case.

Lemma 6.2.5. *Let* $z_0 \in \mathbb{C}_z$ *and suppose that there exist* $p_0 \in S$ *such that* $\pi(p_0) = z_0$ *and* p_0 *is a cluster point for* Σ. *Then* $z_0 \in \Omega$.

Remark 6.2.6. Since we do not know, at this stage, whether Σ is a closed submanifold or not, it is a priori possible that $p_0 \notin \Sigma$. Nevertheless, $\pi^{-1}(z_0) \cap \Sigma \neq \emptyset$.

Proof of Lemma 6.2.5. Let V be a neighborhood of p_0 on which the foliation of $S \cap V$ is trivial. Then, either $\Sigma \cap V$ has finitely many connected components - in this case one of them must contain p_0 - or the connected components of $\Sigma \cap V$ accumulate to the leaf Σ' of $S \cap V$ containing p_0. Then Σ' must be a complex leaf, too. Thus, from Lemma 6.2.1 follows that, if $V' \Subset V$ ($p_0 \in V'$) is small enough, all the leaves of $S \cap V'$ intersect (possibly in V) $\pi^{-1}(z_0)$. By hypothesis

$$V' \cap \Sigma \neq \emptyset,$$

thus Σ contains a leaf of $S \cap V'$, therefore

$$\pi^{-1}(z_0) \cap \Sigma \neq \emptyset. \qquad \square$$

In Section 6.3 we will prove, applying the results on [57], that $\pi|_{\Sigma}^{-1}$ is single-valued. Then, given $z \in \Omega$, we will denote by $w(z)$ (resp. $u(z), v(z)$) the w-coordinates (resp. the u- and v-coordinate) of $\pi|_{\Sigma}^{-1}(z)$. With these notations, we can state the following straightforward corollary of Lemma 6.2.5:

Corollary 6.2.7. *Let* $z_0 \in b\Omega$, *and let* $\{\mathcal{U}_k\}_{k\in\mathbb{N}}$ *be a fundamental system of neighborhoods of* z_0 *in* $\mathbb{C}_z$. *Then, for any* $M > 0$ *there exists* $K \in \mathbb{N}$ *such that* $|w(z)| > M$ *for all* $z \in \Omega \cap \mathcal{U}_k$ *with* $k \geq K$.

Proof. Otherwise, there would exist $M > 0$ and a sequence $\{z_n\}_{n\in\mathbb{N}}$ such that

- $z_n \in \Omega$ for every $n \in \mathbb{N}$;
- $z_n \to z_0$;
- for every $n \in \mathbb{N}$ there exists $p_n \in \Sigma$ such that $\pi(p_n) = z_n$ and $|w(p_n)| \leq M$.

Then $\{p_n\}_{n\in\mathbb{N}}$ would admit an accumulation point p_0 in $\mathbb{C}^2$ such that $\pi(p_0) = z_0$. By Lemma 6.2.5 this would imply $z_0 \in \Omega$, a contradiction. $\qquad \square$

Since, by the main hypothesis, $v(z)$ is bounded on Ω, it follows immediately

Corollary 6.2.8. *Let $z_0 \in b\Omega$, and let $\{\mathcal{U}_k\}_{k \in \mathbb{N}}$ be a fundamental system of neighborhoods of z_0 in $\mathbb{C}_z$. Then for any $M > 0$ there exists $K \in \mathbb{N}$ such that $|u(z)| > M$ for all $z \in \Omega \cap \mathcal{U}_k$ with $k \geq K$.*

Remark 6.2.9. Anyway, since $\Omega \cap \mathcal{U}_k$ need not be connected even for large k, it is possible that u assumes both signs in every neighborhood of z_0 (although later on we are going to prove that it is not the case).

6.3. Analysis of Ω

Our purpose is now to show that Ω is simply connected. In order to achieve this we apply some of the results of [57], in particular the in-depth analysis which is carried out therein on the leaves of the foliation of the Levi-flat solution for graphs. First of all, we prove that $\pi|_{\Sigma}^{-1}$ is actually single-valued over $\Omega = \pi(\Sigma)$.

Lemma 6.3.1. *Let Ω and Σ be as above. Then $\pi|_{\Sigma}^{-1}(z)$ consists of a point for every $z \in \Omega$.*

Proof. Suppose that, for some $z \in \Omega$, there exist $p, q \in \Sigma$ $(p \neq q)$ such that $\pi(p) = \pi(q) = z$. Since, by definition, Σ is connected, there exist an arc $\widetilde{\gamma}$ which joins p and q; let $\gamma = \pi \circ \widetilde{\gamma}$ be the corresponding loop in Ω. Let B be a ball in $\mathbb{C}_z \times \mathbb{R}_u$, centered at z, with a large enough radius such that $\gamma \subset B$ and $\tau \circ \widetilde{\gamma} \subset B$. Then

$$S \cap \tau^{-1}(B) = \Gamma(\rho|_B) \subset \mathbb{C}^2$$

is the Levi flat surface which has the graph

$$S \cap \tau^{-1}(bB) = \Gamma(\rho|_{bB})$$

as boundary. By the results in [57], we conclude that each leaf of the foliation is properly embedded in $S \cap \tau^{-1}(B)$ (observe that, under the hypothesis of Theorem 6.1.2, this fact is granted by our assumption) and, therefore, that $\tau(\Sigma)$ is properly embedded in B. By the choice of B, $\tau(p)$ and $\tau(q)$ belong to the same connected component of $\tau(\Sigma) \cap B$; let Σ' be this component. By Lemma 6.2.1 we have that Σ' is locally a graph over $\mathbb{C}_z$; since B is convex, by Lemma 3.2 in [57] we deduce that Σ' is globally a graph over some subdomain of Ω. Since $\tau(p)$ and $\tau(q)$ have the same projection over Ω, it follows $\tau(p) = \tau(q)$ and consequently $p = q$, a contradiction. $\qquad\square$

By the previous Lemma Σ is represented by the graph of a holomorphic function over Ω. Let us denote by $\mathcal{U}$ (resp. $\mathcal{V}$) the real (resp. the imaginary) part of this function. The following Lemma is an immediate consequence of the results in [57]:

Lemma 6.3.2. Ω *is simply connected.*

Proof. Observe that, if Ω is not simply connected, then $D \cap \Omega$ is not simply connected for some open disc $D \subset \mathbb{C}_z$. Arguing as in the previous Lemma, we prove that $\tau(\Sigma)$ is properly embedded on some subdomain

$$D \times (-R, R) \subset \mathbb{C}_z \times \mathbb{R}_u, \quad R >> 0$$

(again, under the hypothesis of Theorem 6.1.2 this is a direct consequence of the assumption). But $\tau(\Sigma)$ is the graph of $\mathcal{U}$ over $D \cap \Omega$; since $\mathcal{V}$ is a single-valued harmonic conjugate of $\mathcal{U}$, we can apply Lemma 3.3 of [57] and obtain that $D \cap \Omega$ is in fact simply connected. $\square$

6.4. Proof of Theorem 6.1.2

Lemma 6.4.1. *Let C be a connected component of the boundary of Ω. Then there exist a neighborhood $\mathcal{U}$ of C in $\overline{\Omega}$ such that either $u > 0$ on $\mathcal{U}$ or $u < 0$ on $\mathcal{U}$.*

Proof. Let K be a compact connected subset C, chosen in such a way that $C \setminus K$ does not contain relatively compact connected components; it is enough to prove that the thesis holds for any such K. Observe that, since Ω is connected, $C \setminus K$ has at most two connected components. By Corollary 6.2.8, for any $z \in K$ there exist a disc $D(z, \varepsilon)$ such that $|u| > 0$ on $D(z, \varepsilon) \cap \Omega$; thus K can be covered by a finite set $\{D_1, \ldots, D_k\}$ of such discs. If δ is small enough, then

$$\mathcal{U}' = \{z \in \mathbb{C} : d(z, K) < \delta\} \subset D_1 \cup \ldots \cup D_k.$$

The thesis then follows from the fact that there is a connected component of $\mathcal{U}' \cap \Omega$ whose boundary contains K. Suppose that this is not the case, and choose a connected component $\mathcal{V}$ of $\mathcal{U}' \cap \Omega$ such that $E = b\mathcal{V} \cap K \neq \emptyset$. Observe that $b\mathcal{V} = E \cup F \cup G$, where

$$F = b\mathcal{V} \cap \{z \in \mathbb{C} : d(z, K) = \delta\} \text{ and } G = b\mathcal{V} \cap C \setminus K;$$

obviously $E \cap F = \emptyset$ and thus G has at least two connected component. Moreover, E is connected since otherwise $C \setminus K$ would have more than two connected components. But if $E \subsetneq K$ is connected then it can touch at most one connected component of $C \setminus K$ and thus of G; it follows $E = K$. $\square$

Corollary 6.4.2. *Let C be a connected component of $b\Omega$. Then there is a fundamental system $\{\mathcal{V}_n\}_{n \in \mathbb{N}}$ of neighborhoods of C in $\overline{\Omega}$ such that either*

$$\inf_{\mathcal{V}_n} u \to +\infty$$

or

$$\sup_{V_n} u \to -\infty$$

as $n \to \infty$.

Proof. This is a consequence of Corollary 6.2.8 and Lemma 6.4.1.   □

Now we are in position to prove Theorem 6.1.2. Choose a point $w \in C$ and observe that, since Ω is simply connected by Section 6.3, there exists a disc $D = D(w, \varepsilon)$ such that $D\backslash C$ is disconnected and it is not contained in Ω. Let

$$g = \frac{1}{u + iv};$$

then g is well-defined and holomorphic on $D \cap \Omega$. Define a function $\widetilde{g} : D \to \mathbb{C}$ as

$$\widetilde{g}(z) = \begin{cases} g(z), & z \in \overline{\Omega} \cap D; \\ 0, & z \in D \setminus \overline{\Omega}. \end{cases}$$

Then $\widetilde{g}$ is continuous by the previous corollary. Moreover, by definition $\widetilde{g}$ is holomorphic outside the set $\{\widetilde{g} = 0\}$; therefore, by Rado's Theorem, $\widetilde{g} \in \mathcal{O}(D)$. Since $\{\overset{\circ}{\widetilde{g} = 0}\}\neq \emptyset$, we have $\widetilde{g} \equiv 0$ on D and consequently $g \equiv 0$ on Ω, which is a contradiction.

It follows that u cannot be unbounded on Ω. By Corollary 6.2.8 we have that $\Omega = \mathbb{C}_z$ and thus π is onto. Lemma 6.3.1 implies that π is one to one, therefore we can apply Lemma 6.2.2 and conclude that $\Sigma = \{w = c\}$ for some $c \in \mathbb{C}$, whence the thesis of Theorem 6.1.1.

6.5. The result in $\mathbb{C}^n$

The statement of Theorem 6.1.1 can be generalized to the case when S is a Levi-flat hypersurface on $\mathbb{C}^n$. Consider coordinates $(z_1, \ldots, z_{n-1}, w) = (z, w)$, $z_j = x_j + iy_j$, $w = u + iv$, and let $\rho : \mathbb{C}^{n-1} \times \mathbb{R} \to \mathbb{R}$ be a smooth function such that $S = \{v = \rho(z, u)\}$ is a Levi-flat graph. Then we can restate almost verbatim Theorem 6.1.1:

Theorem 6.5.1. *S is foliated by complex hyperplanes, i.e. ρ has the form*

$$\rho(z, u) = \varrho(u)$$

for some smooth function $\varrho : \mathbb{R} \to \mathbb{R}$.

Proof. This is an easy consequence of Theorem 6.1.1. Indeed, let $p_1 = (z_1, u)$ and $p_2 = (z_2, u)$ be two points in $\mathbb{C}_z^{n-1} \times \mathbb{R}_u$ with the same u-coordinate, and consider the complex line $L \subset \mathbb{C}_z^{n-1}$ such that $z_1, z_2 \in L$. Then the restriction of ρ to $L \times \mathbb{R}_u$ has a Levi-flat graph

$$S_L = S \cap (L \times \mathbb{C}_w) \subset L \times \mathbb{C}_w \cong \mathbb{C}^2.$$

Theorem 6.1.1 applies to S_L, showing that $\rho|_{L \times \mathbb{R}_u}$ is a function of u and thus that $\rho(p_1) = \rho(p_2)$. This proves the thesis. $\qquad\square$

6.6. Generalization to a continuous graph

The arguments of the previous sections work in the case that ρ is at least of class C^2. However, it is possible to generalize the result to the case of a continuous graph. In order to achieve this, the Main Theorem of Shcherbina's work [57] (which gives also a description of the leaves of the foliation of the polynomial hull of a graph in $\mathbb{C}^2$) can be applied, rather than Lemmas 3.2 and 3.3. We say that a continuous hypersurface $S \subset \mathbb{C}^n$ (*i.e.* a subset which is locally a graph of a continuous function over an open subset of a real hyperplane of $\mathbb{C}^n$) is Levi-flat if it (locally) separates pseudoconvex domains of $\mathbb{C}^n$. Note that, in the case $n = 2$, S is locally the union of a disjoint family of complex discs (see again [57], Corollary 1.1). So, let $\rho : \mathbb{C}^{n-1} \times \mathbb{R} \to \mathbb{R}$ be a continuous function such that $S = \{v = \rho(z, u)\}$ is a Levi-flat graph; then, as before, we have

Theorem 6.6.1. *S is foliated by complex hyperplanes, i.e.* ρ *has the form*

$$\rho(z, u) = \varrho(u)$$

for some continuous function $\varrho : \mathbb{R} \to \mathbb{R}$.

Once again it is sufficient to show that the statement is true for $n = 2$. In this case we do not know a priori whether S has a foliated atlas; nevertheless, since each $p \in S$ is contained in a germ of holomorphic curve $\Sigma_p \subset S$ (and this germ is unique, see Lemma 4.1 of [57]) we can still consider the maximal connected surface Σ that passes through p. Our aim is to carry out an analysis of Σ similar to the one made in the previous sections for the C^2 case. First of all, we want to generalize Lemma 6.2.1:

Lemma 6.6.2. *Let* Σ *be any leaf of the foliation of S. Then the projection* $\pi|_\Sigma$ *is a local homeomorphism.*

Proof. In this case the fact that $\partial/\partial v \in T(\Sigma)$ does not give a contradiction, since S is only a continuous graph. Instead, we rely on the Main Theorem of [57] Let $p \in \Sigma$, $p = (p_1, p_2)$, let B be a ball in $\mathbb{C}_z \times \mathbb{R}_u$ containing the point $(p_1, \mathrm{Re}\, p_2)$ and consider $\rho|_{bB}$. Then Shcherbina's Theorem applies to $\gamma = \Gamma(\rho|_{bB})$, hence by point (ii) on that Theorem follows that the disc through p is a graph over a domain of $\mathbb{C}_z$. $\qquad\square$

As before, we define $\Omega = \pi(\Sigma)$ and we prove the results corresponding to those in Section 6.3, namely, that $\pi|_\Sigma^{-1}$ is single-valued and that Ω is simply connected.

Lemma 6.6.3. *Let* Ω *and* Σ *be as above. Then* $\pi|_{\Sigma}^{-1}(z)$ *consists of a point for every* $z \in \Omega$.

Proof. We follow the proof of Lemma 6.3.1 and suppose that, for some $z \in \Omega$, there exist $p, q \in \Sigma$ ($p \neq q$) such that $\pi(p) = \pi(q) = z$. We choose an arc $\tilde{\gamma}$ which joins p and q, with $\gamma = \pi \circ \tilde{\gamma}$ the corresponding loop in Ω, and a ball B in $\mathbb{C}_z \times \mathbb{R}_u$, centered at z, with a large enough radius such that $\gamma \subset B$ and $\tau \circ \tilde{\gamma} \subset B$. Then

$$S \cap \tau^{-1}(B) = \Gamma(\rho|_B) \subset \mathbb{C}^2$$

is the Levi-flat surface which has the graph

$$S \cap \tau^{-1}(bB) = \Gamma(\rho|_{bB})$$

as boundary. Since, by Shcherbina's Main Theorem, $S \cap \tau^{-1}(B)$ is the disjoint union of discs which are graphs on $\mathbb{C}_z$, we must have $p = q$, which is a contradiction. $\qquad\square$

Lemma 6.6.4. Ω *is simply connected.*

Proof. As in the previous case, we assume by contradiction that $D \cap \Omega$ is not simply connected for some open disc $D \subset \mathbb{C}_z$. We choose a ball $B \subset \mathbb{C}_z \times \mathbb{R}_u$ such that $B \cap \mathbb{C}_z = D$. Then, by point (ii) of Shcherbina's Main Theorem, the leaves of $S \cap \tau^{-1}(B)$ are graphs over simply connected domains of $\mathbb{C}_z$. It follows that

$$D \cap \Omega = \pi(\Sigma \cap \tau^{-1}(B))$$

must be simply connected. $\qquad\square$

Now we prove the analogous of Lemma 6.2.5:

Lemma 6.6.5. *Let* $z_0 \in \mathbb{C}_z$ *and suppose that there exist* p_0 *such that* $\pi(p_0) = z_0$ *and* p_0 *is a cluster point for* Σ. *Then* $z_0 \in \Omega$.

Proof. In this case we can actually prove that $p_0 \in \Sigma$, *i.e.* Σ is a closed surface. In fact, consider a ball $B \subset \mathbb{C}_z \times \mathbb{R}_u$ which is centered at $\tau(p_0)$. Then $S_B = S \cap \tau^{-1}(B) = \Gamma(\rho|_B)$ is union of disjoint complex discs which are graphs over domains of $\mathbb{C}_z$. Since Σ is a graph over $\mathbb{C}_z$ and contains points of S_B, it must contain exactly one of those discs, which has to be the one passing through p_0 since p_0 is a cluster point. Then $p_0 \in \Sigma$. $\qquad\square$

With the notation adopted in Section 6.2, we then have, with the same proof as 6.2.8, the following

Corollary 6.6.6. *Let* $z_0 \in b\Omega$, *and let* $\{\mathcal{U}_k\}_{k \in \mathbb{N}}$ *be a fundamental system of neighborhoods of* z_0 *in* $\mathbb{C}_z$. *Then for any* $M > 0$ *there exists* $K \in \mathbb{N}$ *such that* $|u(z)| > M$ *for all* $z \in \Omega \cap \mathcal{U}_k$ *with* $k \geq K$.

The rest of the proof of Theorem 6.6.1 goes exactly as in the previous sections.

Chapter 7
Liouville theorems for foliations

In this chapter we want to discuss some results analogous to the ones in Chapter 6, *i.e.* we will show that also in other circumstances (in particular, when the manifold is not a graph) it is possible to conclude that a Levi flat manifold is "trivial", provided that it is bounded in some sense. A first result in this direction is the following:

Theorem 7.0.7. *Let S be a smooth Levi-flat hypersurface of* $\mathbb{C}^n = \mathbb{C}^{n-1} \times \mathbb{C}_w$, *contained in* $C = \{|w| < 1\}$ *and closed in* $\overline{C}$. *Then S is foliated by hyperplanes* $\{w = const.\}$.

In order to treat this problem it is useful a change of perspective, that is, to consider the set S as a whole instead of performing an analysis of every single leaf of its foliation. In other words, we can consider S as an analytic multifunction and then Theorem 7.0.7 becomes a rather easy consequence of the results already obtained for such objects.

7.1. Analytic multifunctions and Liouville theorem

Consider a function $f : \mathbb{C}^n \to \mathcal{P}(\mathbb{C})$, *i.e.* a set-valued function from $\mathbb{C}^n$ to the power set of $\mathbb{C}$. Let $\Gamma(f) \subset \mathbb{C}^{n+1}$ be defined as

$$\Gamma(f) = \bigcup_{z \in \mathbb{C}^n} \{z\} \times f(z).$$

We say that f is an *analytic multifunction* if each value $f(z)$ is a compact set, f is *semicontinuous* (see Chapter 5) and $\mathbb{C}^{n+1} \setminus \Gamma(f)$ is pseudoconvex. With this definition, a holomorphic function $f \in \mathcal{O}(\mathbb{C}^n)$ is clearly an analytic multifunction.

Let $\rho : \mathbb{C}^{n+1} \to \mathbb{R}$ be a continuous plurisubharmonic function. Let $\rho' : \mathbb{C}^n \to \mathbb{R}$ be defined as

$$\rho'(z) = \max_{w \in f(z)} \rho(w).$$

The following result holds ([59]):

Lemma 7.1.1. *For any analytic multifunction f and continuous psh function ρ, ρ' is a plurisubharmonic function.*

Proof. We want to prove that ρ' is subharmonic along every complex line $L \subset \mathbb{C}^n$ *i.e.* it is sufficient to consider the restriction of ρ to $L \times \mathbb{C} \cong \mathbb{C}^2 = \mathbb{C}_z \times \mathbb{C}_w$. Then, let B be any ball on $\mathbb{C}_z$ and consider the harmonic function $h : \overline{B} \to \mathbb{R}$ for which $h|_{bB} = \rho'|_{bB}$. We suppose, by contradiction, that there exist a point $z \in B$ for which $\rho'(z) > h(z)$; defining

$$c = \max\{\rho'(z) - h(z) : z \in \overline{B}\}$$

then $c > 0$. Choose $\zeta_0 \in \mathbb{C}$ such that $\rho'(z_0) - h(z_0) = c$.

Consider the trivial extension of h to $\overline{B} \times \mathbb{C}_w$, $h(z, w) = h(z)$. Then $\rho - h$ is a p.s.h. function and the lower level set $U = \{\rho - h < c\}$ is pseudoconvex. Moreover

$$\Gamma(f)|_{B \times \mathbb{C}_w} \subset U$$

because for every $(z, w) \in \Gamma(f) \cap (B \times \mathbb{C}_w)$ we have that

$$\rho(z, w) - h(z, w) \leq \rho'(z) - h(z) \leq c.$$

But $\Gamma(f)$ touches bU in some point (z_0, w_0) (with z_0 in the internal part of B) and since $\Gamma(f)$ is a pseudoconcave set, this would be a violation of the Kontinuitätsatsatz. □

In [59] it is also proved that the converse of Lemma 7.1.1 holds true (*cf.* Chapter 5). From now on, by analytic multifunction we mean a multifunction for which the conclusion of Lemma 7.1.1 holds true.

The following Liouville result on analytic multifunctions depends only on the property of Lemma 7.1.1.

Lemma 7.1.2. *Let f be an analytic multifunction on $\mathbb{C}^n$, and suppose that f is bounded in the following sense:*

$$\Gamma(f) \subset \{|w| < M\} \subset \mathbb{C}^{n+1}$$

for some $M > 0$. Let $\hat{f}$ be the multifunction defined as

$$\hat{f}(z) = \widehat{f(z)}, \; z \in \mathbb{C}^n$$

where $\widehat{K}$ is the polynomial hull of K. Then $\hat{f}$ is constant.

Proof. Let $P(w)$ be a polynomial on $\mathbb{C}_w$, and denote again by P the trivial extension to $\mathbb{C}^{n+1}$ $P(z, w) = P(z)$. Then $|P|$ is a plurisubharmonic function on $\mathbb{C}^{n+1}$, therefore by Lemma 7.1.1

$$P'(z) = \max\{|P(w)| : w \in f(z)\}$$

is p.s.h. on $\mathbb{C}^n$. But, defining

$$C = \max_{|w| \leq M} P(w)$$

we have that $P'(z) \leq C$ for all $z \in \mathbb{C}^n$. Then, by Liouville's theorem for p.s.h. functions follows that P' is constant. We deduce that $\hat{f}$ is constant. Indeed, in the opposite case we could find $w_1 \in \mathbb{C}$ and $z_1, z_2 \in \mathbb{C}^n$ such that $w_1 \in (\hat{f}(z_1) \setminus \hat{f}(z_2))$, *i.e.* there would exist a polynomial P_1 such that $|P_1(w_1)| > \max_{\hat{f}(z_2)} |P_1|$, hence $P_1'(z_2) < |P_1(w_1)| \leq P_1'(z_2)$ which is a contradiction. □

Example 7.1.3. The hypothesis of Lemma 7.1.2 does not imply that f is in turn a constant multifunction. A simple example is the following:

$$f(z) = \begin{cases} \{|w| = 1\}, & z \neq 0; \\ \{|w| \leq 1\}, & z = 0. \end{cases}$$

Example 7.1.4. A modification of the previous example shows that, even if $\Gamma(f)$ is a (disconnected) manifold, f need not be constant if we adopt the second definition of analytic multifunction (*i.e.* the property discussed in Lemma 7.1.1). In fact, in this case we may define $f(z)$ to be the union of the unit circle bD and any compact set contained in the unit disc D, as any subharmonic function can "detect"the behavior of f only in bD. As we show below, anyway, the result holds if $\Gamma(f)$ has the structure of a (even disconnected) Levi flat manifold (which is obviously not the case in the previous example).

Lemma 7.1.2 provides a tool which allows to prove Theorem 7.0.7 quite easily. In fact, setting

$$f_S(\zeta) = S \cap \{z = \zeta\}$$

for $\zeta \in \mathbb{C}^n$ we have that f_S is by definition an analytic multifunction.

Proof of Theorem 7.0.7. By hypothesis the multifunction f_S is bounded, therefore in view of Liouville's Theorem $\hat{f}_S$ is constant. We have to show that the multifunction f_S is constant, too. In order to do this, choose $z_0 \in \mathbb{C}^n$ in such a way that the complex line $L_{z_0} = \{z = z_0\} \subset \mathbb{C}^{n+1}$

intersects S transversally. This means that $f(z_0)$ is a smooth compact real 1-submanifold of $\mathbb{C}$, *i.e.* a finite set $\{\lambda_i(z_0)\}_{1 \le i \le k(z_0)}$ of simple C^∞ loops contained in $D = \{|w| < 1\}$. Let $U_i(z_0)$ be the bounded connected component of $\mathbb{C} \setminus \lambda_i(z_0)$, and let $\{\alpha_j(z_0)\}_{1 \le j \le h(z_0)}$ be the "maximal"loops, *i.e.* those λ_i's which are not contained in any U_j. For every $z \in \mathbb{C}^n$ for which $L_z \cap S$ is transversal, we make analogous positions and we also define

$$\mathcal{M}(z) = \bigcup_{1 \le i \le h(z)} \alpha_i(z).$$

Let $I \subset \mathbb{C}^n$ be the set of $z \in \mathbb{C}^n$ for which

- L_z has transversal intersection with $\bigcup_\zeta \mathcal{M}(\zeta)$;
- $\mathcal{M}(z) = \mathcal{M}(z_0)$.

It suffices to show that I is both open and closed. Indeed, in this case $\mathbb{C}^n \times \mathcal{M}(z_0)$ has to be an union of connected components of S, from which it follows that $f'_S(z) = f_S(z) \setminus \mathcal{M}(z)$ is an analytic multifunction. Thus, we can prove the statement of 7.0.7 inductively (with the induction performed *e.g.* on the number of loops of $f(z_0)$).

I is open. Let $z_1 \in I$; clearly there exists a neighborhood Ω of z_1 such that $h(z) = h(z_1) \equiv h$ for $z \in \Omega$ and $\mathcal{M}_\Omega = \bigcup_{z \in \Omega} \mathcal{M}(z)$ is a submanifold of $\Omega \times \mathbb{C}_w$ for which $L_z \cap \mathcal{M}_\Omega$ is transversal. Moreover, observe that if $\{V_i(z)\}_{1 \le i \le h}$ are the connected components of $\mathbb{C} \setminus \alpha_i(z)$, then $\hat{f}(z) = \bigcup_i V_i(z)$. This implies immediately that $\mathcal{M}(z)$ is constant on Ω.

I is closed. Let $\mathcal{M}_I = \bigcup_{z \in I} \mathcal{M}_z$ and let $z_2 \in \bar{I}$. Then, $\overline{\mathcal{M}_I} \cap L_{z_2} = \mathcal{M}(z_0)$; moreover, since S is a smooth manifold, we have

$$T_{(z_2, w')}(S) \supset \{(z, w) \in \mathbb{C}^{n+1} : w = w'\}$$

for every $w' \in f(z_0)$. But, since we clearly have $\overline{\mathcal{M}_I} \cap L_{z_2} = \mathcal{M}(z_2)$, this implies that $L_{z_2} \cap \mathcal{M}$ is transversal, *i.e.* $z_2 \in I$. $\square$

7.2. Higher codimension

The analogous of Theorem 7.0.7 for Levi flat surfaces S of higher codimension can also be proved with the methods of analytic multifunctions. However, analyticity of them multifunction f_S defined by S is more involved, due to the fact that S is no longer pseudoconvex. Also the proof of the fact that f_S is constant whenever $\hat{f}_S$ is constant needs to be adapted using [60] (see the proof of Theorem 7.2.1).

We consider a real $(2d - 1)$-codimensional submanifold $S \subset \mathbb{C}^{n+d} \cong \mathbb{C}^n \times \mathbb{C}^d$, with coordinates $z_1, \ldots, z_n, w_1, \ldots, w_d$.

Theorem 7.2.1. *Let S be a $(2d - 1)$-codimensional closed, Levi flat submanifold (i.e. foliated by complex leaves of dimension n) of $\mathbb{C}^{n+d}$, contained in*

$$C = \{(z, w) \in \mathbb{C}^{n+d} : \sum_{i=1}^{d} |w_i|^2 < 1\}.$$

Then S is foliated by complex n-planes of the kind $\{w_1 = c_1, \ldots, w_d = c_d\}$.

Let $f : \mathbb{C}^n \to \mathcal{P}(\mathbb{C}^d)$ be a function from $\mathbb{C}^n$ to the subsets of $\mathbb{C}^d$, $d \geq 2$. We recall that, according to the definition we are adopting, f is an analytic multifunction if $f(z)$ is compact for each $z \in \mathbb{C}^n$ and, for every continuous plurisubharmonic function $\rho : \mathbb{C}^{n+d} \to \mathbb{R}$, the function $\rho' : \mathbb{C}^n \to \mathbb{R}$ defined as

$$\rho'(z) = \max_{f(z)} \rho(z, w)$$

is plurisubharmonic.

Let L_z, $z \in \mathbb{C}^n$, be the vertical complex d-plane over z i.e.

$$L_z = \{(z, w) : w \in \mathbb{C}^d\}.$$

Consider the set-valued function f_S defined by $f_S(z) = L_z \cap S$ (generically, $f_S(z)$ is the union of a finite number of loops). We want to show that f_S is an analytic multifunction.

Lemma 7.2.2. *f_S is an analytic multifunction.*

Proof. Let $\rho : \mathbb{C}^{n+d} \to \mathbb{R}$ be a p.s.h. function, and define ρ' as above. Let $z_0 \in \mathbb{C}^n$, and let $\mathcal{L} \subset \mathbb{C}^n$ be a complex line passing through z_0. For a generic choice of $\mathcal{L}$, the intersection of S with the complex $(d + 1)$-plane

$$\{(z, w) \in \mathbb{C}^{n+d} : z \in \mathcal{L}\}$$

is transversal, and thus a Levi flat submanifold of $\mathbb{C}^{d+1}$. Therefore, since it is sufficient to show that the restriction of ρ' to a generic $\mathcal{L}$ is subharmonic, we can suppose $n = 1$.

Assume, then, that f_S is a $\mathcal{P}(\mathbb{C}^d)$-valued multifunction defined over $\mathbb{C}_z$, and fix $z_0 \in \mathbb{C}$. If $w \in f(z_0)$, we denote by Σ_w the leaf of the foliation of S through w. Two cases are possible:

- $T_{(z_0, w)}(\Sigma_w) \not\subseteq \mathbb{C}^d$.

  In this case, for a sufficiently small neighborhood $V_w = (\Delta \times U)_w$ of (z_0, w) we have that $\Sigma_w \cap V_w$ can be written as

  $$\Sigma_w \cap V_w = \{(z, w) \in \Delta \times U : w_1 = g_1^w(z), \ldots, w_d = g_d^w(z)\}$$

for some holomorphic function $g_i^w \in \mathcal{O}(\Delta)$. Moreover, observe that for $w' \in f(z_0)$ in a small enough neighborhood $\mathcal{W}_w$ of w, we can choose a Δ which does not depend on w';

- $T_{(z_0,w)}(\Sigma_w) \subset \mathbb{C}^d$.

In this case, consider the restriction of the projection $\pi : \mathbb{C}^{d+1} \to \mathbb{C}$ to a small neighborhood $\mathcal{V}_w$ of (z_0, w) in Σ_w. We can suppose that $\mathcal{V}_w$ is a local chart such that $(z_0, w) \cong 0$. Denote by ζ the complex coordinate on $\mathcal{V}_w$. Since $\pi|_{\mathcal{V}_w}$ is a holomorphic function, and its prime derivative vanishes in 0, there exists $k \geq 1$ such that

$$\frac{\partial^k}{\partial \zeta}\pi|_{\mathcal{V}_w} = 0, \quad \frac{\partial^{k+1}}{\partial \zeta}\pi|_{\mathcal{V}_w} \neq 0.$$

Otherwise, we would have $\pi|_{\mathcal{V}_w} \equiv z_0$ and thus Σ_w would be a complex line contained in $\mathbb{C}^d$, which is impossible since it must be contained in the cylinder C of Theorem 7.2.1. It follows that $\pi|_{\mathcal{V}_w}$ is a k-sheeted covering over some neighborhood Δ of z_0. Now, the restriction of π to the leaves $\Sigma_{w'}$ passing through the points (z_0, w') of a small neighborhood of (z_0, w) can be interpreted as a smooth one-parameter family of holomorphic functions $\pi_t : \mathcal{V}_k \to \mathbb{C}_z$, such that $\pi_0 = \pi$. For $|t| << 1$, the argument principle implies that the sum of the orders of the zeroes of $(\partial/\partial\zeta)\pi_t$ is still $k - 1$. This in turn means that for w' sufficiently close to w the projection $\pi|_{\Sigma_{w'}}$ is still a k-sheeted covering over some neighborhood $\Delta_{w'}$; in a possibly smaller neighborhood $\mathcal{W}_w$ we can assume to have chosen a Δ independent from w'.

Since $f(z_0)$ is a compact set, we can choose finitely many open sets as above, $\mathcal{W}_{w_1}, \ldots, \mathcal{W}_{w_h}$, in such a way that

$$\bigcup_{i=1}^{h} \mathcal{W}_{w_i} = f(z_0).$$

Choose a disc $\Delta \subset \Delta_{w_1} \cap \ldots \cap \Delta_{w_h}$. We claim that ρ' is plurisubharmonic on Δ. In order to prove this, choose $w \in f(z_0)$:

- if $w \in \mathcal{W}_{w_j}$ with w_j of the first kind, then we define

$$\rho_w^j = \rho|_{\Sigma_w \cap \pi^{-1}(\Delta)};$$

- if $w \in \mathcal{W}_{w_j}$ with w_j of the second kind, we define

$$\rho_w^j = (\max_{\Sigma_w \cap \pi^{-1}(\Delta_{w_j})} \rho(z, w))|_\Delta.$$

In both cases, ρ_w^j is a psh function. Observe that possibly $\rho_w^i \neq \rho_w^j$ when $i \neq j$. Nevertheless, consider

$$\varrho(z) = \max_{1 \leq i \leq h, w \in f(z_0)} \rho_w^i; \qquad (7.2.1)$$

we have that $\varrho(z) = \rho'(z)$. In fact, the arguments above show that

$$\bigcup_{w \in f(z_0)} \Sigma_w \cap \pi^{-1}(\Delta) = S \cap \pi^{-1}(\Delta)$$

and thus the maximum of equation (7.2.1) is performed exactly on $f(z)$ rather than a proper subset (as would be the case if leaves of S which accumulate on $f(z_0)$ without intersecting it existed). Since we already know that $\rho'(z)$ is continuous, (7.2.1) implies that $\rho'(z)$ is plurisubharmonic. $\qquad \square$

Lemma 7.2.2 allows to prove, exactly in the same way as before, that $\hat{f}_S$ is a constant multifunction. We have to show, again, that this fact forces f_S to be constant.

Proof of Theorem 7.2.1. Observe that, for z belonging to a dense, open subset J of $\mathbb{C}^n$, L_z intersects S transversally. For $z \in J$, $f(z) = L_z \cap S$ is the disjoint union of a finite set $\{\gamma_i(z)\}_{1 \leq i \leq k(z)}$ of loops in $\mathbb{C}^d$. It is a well-known fact ([60]) that, in this case, the polynomial hull $\hat{f}(z)$ of $f(z)$ is given by the union of some of the loops γ_i and some complex varieties Λ_j whose boundaries are the others γ_i's. We choose the minimal subsets of loops $\{\alpha_i(z)\}_{1 \leq i \leq h(z)}$ such that, if $\mathcal{M}(z) = \alpha_1 \cup \ldots \cup \alpha_{h(z)}$, then $\widehat{\mathcal{M}}(z) = \hat{f}(z)$; observe that $\mathcal{M}(z)$ is univocally defined. It is sufficient to prove that $\mathcal{M}(z)$ is constant, because in such a case we can proceed inductively as in the proof of Theorem 7.0.7. Because of the structure of the hull $\hat{f}_S(z)$, it is clear that $\mathcal{M}(z)$ is constant for $z \in J$. Moreover, arguing again as in 7.0.7, it is clear that

$$\mathcal{M} = \bigcup_z \mathcal{M}(z)$$

is a manifold, which has transversal intersection with L_z also for $z \in \overline{J}$. It follows that $\mathcal{M}$, and therefore f_S, is constant. $\qquad \square$

Bibliography

[1] H. ALEXANDER, *A note on polynomial hulls*, Proc. Amer. Math. Soc. **33** (1972), 389–391.

[2] H. ALEXANDER and J. WERMER, *Polynomial hulls with convex fibers*, Math. Ann. **271** (1985), no. 1, 99–109. MR **779607** (**86i:**32025)

[3] H. ALEXANDER and J. WERMER, *Envelopes of holomorphy and polynomial hulls*, Math. Ann. **281** (1988), no. 1, 13–22. MR **944599** (**89i:**32026)

[4] A. ANDREOTTI and C. D. HILL, *E. E. Levi convexity and the Hans Lewy problem. I. Reduction to vanishing Theorems*, Ann. Scuola Norm. Sup. Pisa Cl. Sci. (3) **26** (1972), 325–363.MR0460725 (57 #718)

[5] A. ANDREOTTI and C. D. HILL, *E. E. Levi convexity and the Hans Lewy problem. II. Vanishing Theorems*, Ann. Scuola Norm. Sup. Pisa Cl. Sci. (3) **26** (1972), 747–806.MR0477150 (57 #16693)

[6] A. ANDREOTTI and Yum-tong Siu, *Projective embedding of pseudoconcave spaces*, Ann. Scuola Norm. Sup. Pisa Cl. Sci. (3) **24** (1970), 231–278.

[7] B. AUPETIT and A. ZRAÏBI, *Distribution des valeurs des fonctions analytiques multiformes*, Studia Math. **79** (1984), no. 3, 217–226 (French, with English summary). MR **781719** (**86m:**46046)

[8] D. E. BARRETT and T. INABA, *On the topology of compact smooth three-dimensional Levi-flat hypersurfaces*, J. Geom. Anal. **2** (1992), no. 6, 489–497. MR **1189041** (**94e:**32034)

[9] E. BEDFORD and B. GAVEAU, *Envelopes of holomorphy of certain 2-spheres in* $\mathbf{C}^2$, Amer. J. Math. **105** (1983), no. 4, 975–1009.MR708370 (84k:32016)

[10] E. BEDFORD and W. KLINGENBERG, *On the envelope of holomor-*

phy of a 2-sphere in $\mathbf{C}^2$, J. Amer. Math. Soc. **4** (1991), no. 3, 623–646.MR1094437 (92j:32034)

[11] E. BISHOP, *Differentiable manifolds in complex Euclidean space*, Duke Math. J. **32** (1965), 1–21.MR0200476 (34 #369)

[12] S. BOCHNER, *Analytic and meromorphic continuation by means of Green's formula*, Ann. of Math. (2) **44** (1943), 652–673. MR 0009206 (5,116f)

[13] A. BOGGESS, *CR manifolds and the tangential Cauchy-Riemann complex*, Studies in Advanced Mathematics, CRC Press, Boca Raton, FL, 1991. MR **1211412 (94e:**32035)

[14] R. L. BRYANT, *Levi-flat minimal hypersurfaces in two-dimensional complex space forms*, hundred years after Sophus Lie (Kyoto/Nara, 1999), Adv. Stud. Pure Math., vol. 37, Math. Soc. Japan, Tokyo, 2002, pp. 1–44. MR **1980895 (2005a:**53099)

[15] A. CANDEL and L. CONLON, *Foliations. I*, Graduate Studies in Mathematics, vol. 23, American Mathematical Society, Providence, RI, 2000. MR **1732868 (2002f:**57058)

[16] G. DELLA SALA and A. SARACCO, *Non-compact boundaries of complex analytic varieties*, Internat. J. Math. **18** (2007), no. 2, 203–218. MR 2307422

[17] G. DELLA SALA and A. SARACCO, *Semi-local exten sion of maximally complex submanifolds*, preprint (arXiv: math/0607747 [math.CV]) (2006).

[18] T.-C. DINH, *Conjecture de Globevnik-Stout et théorème de Morera pur une chaîne holomorphe*, Ann. Fac. Sci. Toulouse Math. (6) **8** (1999), no. 2, 235–257 (French, with English and French summaries). MR **1751442 (2001a:**32014)

[19] P. DOLBEAULT, *Sur les chaînes maximalement complexes de bord donné*, Geometric measure theory and the calculus of variations (Arcata, Calif., 1984), Proc. Sympos. Pure Math., vol. 44, Amer. Math. Soc., Providence, RI, 1986, pp. 171–205 (French). MR **840273 (87m:**32023)

[20] P. DOLBEAULT and G. HENKIN, *Surfaces de Riemann de bord donné dans* $\mathbf{CP}^n$, Contributions to complex analysis and analytic geometry, Aspects Math., E26, Vieweg, Braunschweig, 1994, pp. 163–187 (French). MR **1319348 (96a:**32020)

[21] P. DOLBEAULT and G. HENKIN, *Chaînes holomorphes de bord donné dans* $\mathbf{CP}^n$, Bull. Soc. Math. France **125** (1997), no. 3, 383–445 (French, with English and French summaries). MR **1605457 (98m:**32014)

[22] P. DOLBEAULT, G. TOMASSINI, and D. ZAITSEV, *On boundaries of Levi-flat hypersurfaces in* $\mathbb{C}^n$, C. R. Math. Acad. Sci. Paris **341** (2005), no. 6, 343–348 (English, with English and French summaries). MR **2169149 (2006e:**32048)

[23] Y. ELIASHBERG, *Filling by holomorphic discs and its applications*, Geometry of low-dimensional manifolds, 2 (Durham, 1989), London Math. Soc. Lecture Note Ser., vol. 151, Cambridge Univ. Press, Cambridge, 1990, pp. 45–67. MR **1171908 (93g:**53060)

[24] H. FEDERER, "Geometric Measure Theory", Die Grundlehren der mathematischen Wissenschaften, Band 153, Springer-Verlag New York Inc., New York, 1969. MR 0257325 (41 #1976)

[25] F. FORSTNERIČ, *On the boundary regularity of proper mappings*, Ann. Scuola Norm. Sup. Pisa Cl. Sci. (4) **13** (1986), no. 1, 109–128. MR **863637 (87m:**32055)

[26] F. FORSTNERIČ, *Polynomial hulls of sets fibered over the circle*, Indiana Univ. Math. J. **37** (1988), no. 4, 869–889. MR **982834 (90g:**32018)

[27] F. FORSTNERIČ and C. LAURENT-THIÉBAUT, *Stein compacts in Levi-flat hypersurfaces*, Trans. Amer. Math. Soc. **360** (2008), no. 1, 307–329 (electronic). MR 2342004

[28] M. P. GAMBARYAN, *Regularity condition for complex films*, Uspekhi Mat. Nauk **40** (1985), no. 3(243), 203–204 (Russian). MR **795194 (86m:**32031)

[29] F. R. HARVEY and H. B. LAWSON JR., *On boundaries of complex analytic varieties. I*, Ann. of Math. (2) **102** (1975), no. 2, 223–290.MR0425173 (54 #13130)

[30] F. R. HARVEY and H. B. LAWSON JR., *Addendum to Theorem 10.4 in "Boundaries of analytic varieties"*, [arXiv: math.CV/0002195] (2000).

[31] F. R. HARVEY and H. B. LAWSON JR., *On boundaries of complex analytic varieties. II*, Ann. Math. (2) **106** (1977), no. 2, 213–238. MR 0499285 (58 #17186)

[32] F. R. HARVEY and H. B. LAWSON JR., *Boundaries of varieties in projective manifolds*, J. Geom. Anal. **14** (2004), no. 4, 673–695.

[33] F. R. HARVEY and H. B. LAWSON JR., *Projective Linking and Boundaries of Positive Holomorphic Chains in Projective Manifolds*, [arXiv: math.CV/0512379] (2005).

[34] F. R. HARVEY and H. B. LAWSON JR., *Projective Linking and Boundaries of Positive Holomorphic Chains in Projective Manifolds, Part II*, [arXiv: math.CV/0608029] (2006).

[35] F. R. HARVEY and H. B. LAWSON JR., *Boundaries of Positive Holomorphic Chains*, [arXiv: math.CV/0610533] (2006).

[36] L. HÖRMANDER, "An Introduction to Complex Analysis in Several Variables", D. Van Nostrand Co., Inc., Princeton, N.J.-Toronto, Ont.-London, 1966. MR 0203075 (34 #2933)

[37] C. E. KENIG and S. M. WEBSTER, *The local hull of holomorphy of a surface in the space of two complex variables*, Invent. Math. **67** (1982), no. 1, 1–21. MR **664323 (84c:**32014)

[38] H. KNESER, *Die Randwerte eine analytischen Funktion zweier Veränderlichen*, Monatsh. Math. Phys. **43** (1936), 364–380 (German).

[39] J. J. KOHN and H. ROSSI, *On the extension of holomorphic functions from the boundary of a complex manifold*, Ann. of Math. (2) **81** (1965), 451–472. MR 0177135 (31 #1399)

[40] C. LAURENT-THIÉBAUT, *Sur l'équation de Cauchy-Riemann tangentielle dans une calotte strictement pseudoconvexe*, Internat. J. Math. **16** (2005), no. 9, 1063–1079 (French, with English and French summaries). MR **2180065 (2006m:**32044)

[41] E. E. LEVI, *Studio sui punti singolari essenziali delle funzioni analitiche di due o più variabili complesse*, Ann. Mat. Pura Appl. **17** (1910), 61–88 (Italian).

[42] H. LEWY, *On the local character of the solutions of an atypical linear differential equation in three variables and a related Theorem for regular functions of two complex variables*, Ann. of Math. (2) **64** (1956), 514–522. MR 0081952 (18,473b)

[43] H. LEWY, *On the relations governing the boundary values of analytic functions of two complex variables*, Comm. Pure Appl. Math. **9** (1956), 295–297. MR 0079099 (18,28e)

[44] G. LUPACCIOLU and Giuseppe Tomassini, *An extension Theorem for CR-functions*, Ann. Mat. Pura Appl. (4) **137** (1984), 257–263 (Italian, with English summary).MR772261 (86e:32021)

[45] G. LUPACCIOLU, *A Theorem on holomorphic extension of CR-functions*, Pacific J. Math. **124** (1986), no. 1, 177–191.MR850675 (87k:32026)

[46] G. LUPACCIOLU, *Valeurs au bord de fonctions holomorphes dans des domaines non bornés de* $\mathbf{C}^n$, C. R. Acad. Sci. Paris Sér. I Math. **304** (1987), no. 3, 67–69 (French, with English summary). MR **878828 (88c:**32024)

[47] E. MARTINELLI, *Sulla determinazione di una funzione analitica di più variabili complesse in un campo, assegnatane la traccia sulla frontiera*, Ann. Mat. Pura Appl. (4) **55** (1961), 191–202 (Italian). MR 0170032 (30 #273)

[48] A. NEWLANDER and L. NIRENBERG, *Complex analytic coordinates in almost complex manifolds*, Ann. of Math. (2) **65** (1957), 391–404. MR 0088770 (19,577a)

[49] K. OKA, *Note sur les familles de fonctions analytiques multiformes*, J. Sci. Hiroshima Univ. Ser. A **4** (1934), 93–98.

[50] C. PARRINI and G. TOMASSINI, $\bar{\partial}\mu = f$: *existence of bounded solutions in unbounded domains*, Boll. Un. Mat. Ital. B (7) **1** (1987), no. 4, 1211–1226 (Italian, with English summary). MR **923449** (**88m**:32038)

[51] T. RANSFORD, *A new approach to analytic multifunctions*, Set-Valued Anal. **7** (1999), no. 2, 159–194. MR **1716030 (2000j**:30081)

[52] C. REA, *Extension holomorphe bilatérale des fonctions CR données sur une hypersurface différentiable de* $\mathbf{C}^2$, C. R. Acad. Sci. Paris Sér. I Math. **294** (1982), no. 17, 577–579 (French, with English summary). MR **663083 (83f**:32018)

[53] C. REA, *Prolongement holomorphe des fonctions CR, conditions suffisantes*, C. R. Acad. Sci. Paris Sér. I Math. **297** (1983), no. 3, 163–166 (French, with English summary). MR **725396 (85a**:32021)

[54] A. SARACCO and G. TOMASSINI, *Cohomology and extension problems for semi q-coronae*, Math. Z. **256** (2007), no. 4, 737–748. MR 2308887

[55] A. SARACCO and G. TOMASSINI, *Cohomology of semi q-coronae and extension of analytic subsets*, preprint (2007).

[56] F. SEVERI, *Contributi alla teoria delle funzioni biarmoniche*, Mem. R. Acc. d'Italia **2** (1931) (Italian).

[57] N. V. SHCHERBINA, *On the polynomial hull of a graph*, Indiana Univ. Math. J. **42** (1993), no. 2, 477–503.MR1237056 (95e:32017)

[58] N. V. SHCHERBINA and G. TOMASSINI, *The Dirichlet problem for Levi-flat graphs over unbounded domains*, Internat. Math. Res. Notices (1999), no. 3, 111–151.MR1672246 (2000a:32018)

[59] Z. SLODKOWSKI, *Analytic set-valued functions and spectra*, Math. Ann. **256** (1981), no. 3, 363–386. MR **626955 (83b**:46070)

[60] G. STOLZENBERG, *Uniform approximation on smooth curves*, Acta Math. **115** (1966), 185–198. MR 0192080 (33 #307)

[61] W. P. THURSTON, *A generalization of the Reeb stability theorem*, Topology **13** (1974), 347–352. MR 0356087 (50 #8558)

[62] G. TOMASSINI, *Tracce delle funzioni olomorfe sulle sottovarietà analitiche reali d'una varietà complessa*, Ann. Scuola Norm. Sup. Pisa Cl. Sci. (3) **20** (1966), 31–43 (Italian). MR 0206992 (34 #6808)

[63] G. TOMASSINI, *Sur les algèbres $A^0(\bar{D})$ et $A^\infty(\bar{D})$ d'un domaine pseudoconvexe non borné*, Ann. Scuola Norm. Sup. Pisa Cl. Sci. (4) **10** (1983), no. 2, 243–256 (French).

[64] J.-M. TRÉPREAU, *Sur le prolongement holomorphe des fonctions C-R définies sur une hypersurface réelle de classe C^2 dans $\mathbf{C}^n$*, Invent. Math. **83** (1986), no. 3, 583–592 (French).MR827369 (87f:32035)

[65] A. E. TUMANOV, *Extension of CR-functions into a wedge from a manifold of finite type*, Mat. Sb. (N.S.) **136(178)** (1988), no. 1, 128–139 (Russian); English transl., Math. USSR-Sb. **64** (1989), no. 1, 129–140. MR **945904 (89m:**32027)

[66] J. WERMER, *The hull of a curve in C^n*, Ann. of Math. (2) **68** (1958), 550–561.MR0100102 (20 #6536)

THESES

This series gathers a selection of outstanding Ph.D. theses defended at the Scuola Normale Superiore since 1992.

Published volumes

1. F. COSTANTINO, *Shadows and Branched Shadows of 3 and 4-Manifolds*, 2005. ISBN 88-7642-154-8

2. S. FRANCAVIGLIA, *Hyperbolicity Equations for Cusped 3-Manifolds and Volume-Rigidity of Representations*, 2005. ISBN 88-7642-167-x

3. E. SINIBALDI, *Implicit Preconditioned Numerical Schemes for the Simulation of Three-Dimensional Barotropic Flows*, 2007.
ISBN 978-88-7642-310-9

4. F. SANTAMBROGIO, *Variational Problems in Transport Theory with Mass Concentration*, 2007. ISBN 978-88-7642-312-3

5. M. R. BAKHTIARI, *Quantum Gases in Quasi-One-Dimensional Arrays*, 2007. ISBN 978-88-7642-319-2

6. T. SERVI, *On the First-Order Theory of Real Exponentiation*, 2008.
ISBN 978-88-7642-325-3

7. D. VITTONE, *Submanifolds in Carnot Groups*, 2008.
ISBN 978-88-7642-327-7

8. A. FIGALLI, *Optimal Transportation and Action-Minimizing Measures*, 2008. ISBN 978-88-7642-330-7

9. A. SARACCO, *Extension Problems in Complex and CR-Geometry*, 2008. ISBN 978-88-7642-338-3

10. L. MANCA, *Kolmogorov Operators in Spaces of Continuous Functions and Equations for Measures*, 2008. ISBN 978-88-7642-336-9

11. M. LELLI, *Solution Structure and Solution Dynamics in Chiral Ytter-bium(III) Complexes*, 2009. ISBN 978-88-7642-349-9

12. G. CRIPPA, *The Flow Associated to Weakly Differentiable Vector Fields*, 2009. ISBN 978-88-7642-340-6

13. F. CALLEGARO, *Cohomology of Finite and Affine Type Artin Groups over Abelian Representations*, 2009. ISBN 978-88-7642-345-1

14. G. DELLA SALA, *Geometric Properties of Non-compact CR Manifolds*, 2009. ISBN 978-88-7642-348-2

Volumes published earlier

H. Y. FUJITA, *Equations de Navier-Stokes stochastiques non homogènes et applications*, 1992.

G. GAMBERINI, *The minimal supersymmetric standard model and its phenomenological implications*, 1993. ISBN 978-88-7642-274-4

C. DE FABRITIIS, *Actions of Holomorphic Maps on Spaces of Holomorphic Functions*, 1994. ISBN 978-88-7642-275-1

C. PETRONIO, *Standard Spines and 3-Manifolds*, 1995.
ISBN 978-88-7642-256-0

I. DAMIANI, *Untwisted Affine Quantum Algebras: the Highest Coefficient of* det H_η *and the Center at Odd Roots of 1*, 1996.
ISBN 978-88-7642-285-0

M. MANETTI, *Degenerations of Algebraic Surfaces and Applications to Moduli Problems*, 1996. ISBN 978-88-7642-277-5

F. CEI, *Search for Neutrinos from Stellar Gravitational Collapse with the MACRO Experiment at Gran Sasso*, 1996. ISBN 978-88-7642-284-3

A. SHLAPUNOV, *Green's Integrals and Their Applications to Elliptic Systems*, 1996. ISBN 978-88-7642-270-6

R. TAURASO, *Periodic Points for Expanding Maps and for Their Extensions*, 1996. ISBN 978-88-7642-271-3

Y. BOZZI, *A study on the activity-dependent expression of neurotrophic factors in the rat visual system*, 1997. ISBN 978-88-7642-272-0

M. L. CHIOFALO, *Screening effects in bipolaron theory and high-temperature superconductivity*, 1997. ISBN 978-88-7642-279-9

D. M. CARLUCCI, *On Spin Glass Theory Beyond Mean Field*, 1998.
ISBN 978-88-7642-276-8

G. LENZI, *The MU-calculus and the Hierarchy Problem*, 1998.
ISBN 978-88-7642-283-6

R. SCOGNAMILLO, *Principal G-bundles and abelian varieties: the Hitchin system*, 1998. ISBN 978-88-7642-281-2

G. ASCOLI, *Biochemical and spectroscopic characterization of CP20, a protein involved in synaptic plasticity mechanism*, 1998.
ISBN 978-88-7642-273-7

F. PISTOLESI, *Evolution from BCS Superconductivity to Bose-Einstein Condensation and Infrared Behavior of the Bosonic Limit*, 1998.
ISBN 978-88-7642-282-9

L. PILO, *Chern-Simons Field Theory and Invariants of 3-Manifolds*,1999. ISBN 978-88-7642-278-2

P. ASCHIERI, *On the Geometry of Inhomogeneous Quantum Groups*, 1999. ISBN 978-88-7642-261-4

S. CONTI, *Ground state properties and excitation spectrum of correlated electron systems*, 1999. ISBN 978-88-7642-269-0

G. GAIFFI, *De Concini-Procesi models of arrangements and symmetric group actions*, 1999. ISBN 978-88-7642-289-8

N. DONATO, *Search for neutrino oscillations in a long baseline experiment at the Chooz nuclear reactors*, 1999. ISBN 978-88-7642-288-1

R. CHIRIVÌ, *LS algebras and Schubert varieties*, 2003. ISBN 978-88-7642-287-4

V. MAGNANI, *Elements of Geometric Measure Theory on Sub-Riemannian Groups*, 2003. ISBN 88-7642•152-1

F. M. ROSSI, *A Study on Nerve Growth Factor (NGF) Receptor Expression in the Rat Visual Cortex: Possible Sites and Mechanisms of NGF Action in Cortical Plasticity*, 2004. ISBN 978-88-7642-280-5

G. PINTACUDA, *NMR and NIR-CD of Lanthanide Complexes*, 2004. ISBN 88-7642-143-2

Fotocomposizione "CompoMat" Loc. Braccone, 02040 Configni (RI) Italy
Finito di stampare nel mese di dicembre 2009
dalla CSR srl, Via di Pietralata, 157, 00158 Roma